Chef's Guide
to
CHARCUTERIE

CRC Press
Taylor & Francis Group
Boca Raton   London   New York

CRC Press is an imprint of the
Taylor & Francis Group, an **informa** business

Chef's Guide to
to
CHARCUTERIE

Jacques Brevery

CRC Press
Taylor & Francis Group
Boca Raton   London   New York

CRC Press is an imprint of the
Taylor & Francis Group, an **informa** business

CRC Press
Taylor & Francis Group
6000 Broken Sound Parkway NW, Suite 300
Boca Raton, FL 33487-2742

© 2013 by Taylor & Francis Group, LLC
CRC Press is an imprint of Taylor & Francis Group, an Informa business

No claim to original U.S. Government works

Printed in the United States of America on acid-free paper
Version Date: 20121031

International Standard Book Number: 978-1-4665-5984-4 (Hardback)

Visit the Taylor & Francis Web site at
http://www.taylorandfrancis.com

and the CRC Press Web site at
http://www.crcpress.com

CONTENTS

FOREWORD

Sometimes, when I'm flipping through the cookbooks in the bookstores, I'm left with a gnawing feeling there are certain things about cooking that needed to be expressed—a sense of something missing, something left unexplained, like the frustration of being in a discussion at a dinner party and feeling unable to get an important point across. Chef Jacques Brevery, CEC, AAC is now taking some of that frustration out of our lives with this book on charcuterie.

Chef Jacques decided to create a kind of guidebook or reference of techniques for people who may be used to charcuterie as part of the garde manager chef's repertoire, but who want more confidence and sense of freedom in the kitchen. This branch of cooking devoted to prepared meat products such as bacon, ham, sausage, terrines, galantines, patés, and confit show a noble history. First recorded back to the first century AD, Romans regulated the imports of salted meat, but in the 15th century the French opened this market through their local guilds, producing charcuterie most varied and sometimes distinctively different from region to region, truly revolutionized these techniques.

Despite my training in classical cooking techniques, I have long valued the skillsets great cookbooks can offer, and if you understand the logic of how these techniques work, you'll be able to improvise intuitively and give your own special style and identity to your cooking. This is invaluable to the competition chef; special attention should be given to chapters that include the fundamental "how-to" techniques that show and tell you right off the bat what to do.

I am delighted that Chef Jacques Brevery has put together this comprehensive book of charcuterie that is certain to become a treasured resource for anyone who cooks, both professionally and at home. It has everything needed to truly understand and master a wide array of charcuterie from around the world, especially French, which are now such an important part of the American dining landscape. The detailed and vivid descriptions of the history of the food culture for each item are not only fascinating, but also give the reader an understanding of how each element evolved.

I know Jacques's instructions of charcuterie are certain to have a prominent place in my culinary library, and I hope yours as well.

Stafford T. DeCambra, CEC, CCE, CCA, AAC
Corporate Executive Chef
PCI Gaming, Atmore, Alabama
American Academy of Chefs—Chairman

PREFACE

This book is intended to serve as a guide for how to transform lesser quality meats and organ meats into enjoyable and beautiful foods, which is the art of professional charcuterie.

This is a complete collection of recipes and information from my career as a chef, much of which I learned from these great men with whom I have crossed paths over the years, including:

Dr. Jean-Claude Frentz, noted French author and expert in charcuterie. He wrote many books on charcutery, like *The encyclopedia of Charcutery (soussana)*, *The Compagnon charcutier*, and *la charcuterie in simplicity in Quebec–Canada*.

Mr. Andre Coquemont, Master-Charcutier from Brittany, France. I had the opportunity to train with him at seminars and attend his lectures in Belgium.

Professor Robert Fagneray, my mentor and teaching chair of the Technical Institute of Arts and Crafts (IATA) in Namur, Belgium, who also own a very nice shop in Neufchateau-belgium.

These three *Compagnons charcutiers* were my teachers over the past forty years. They are the best in their craft, and I am fortunate to have had the opportunity to learn from them.

I hope this book will open your mind to the fine art of charcuterie and teach you how to produce and display these products in a professional way. I intend for this book to be a reference for young chefs, providing them with accurate recipes for classic preparations as well as new ideas that will allow them to expand and improve their portfolio of recipes.

Nowadays, as technology is evolving so rapidly, charcuterie requires more than just a daily performance of routine tasks. Chefs must understand why and how these traditional charcuterie processes work. I am convinced my guide will provide this enrichment for culinary professionals, and I wish success for all of my readers with their cooking.

Acknowledgments

During the realization of this present notebook on Charcuterie-food processing, I was provided with the help of McFatter Technical Center, where I have taught for over 3 years, in Davie, Florida.

I also want to point out the effective assistance and friendship I've been granted from J. Claude Frentz, Master Charcutier, former director of the Charcuterie Soussana, S.A., Institute of Paris; also Andre Coquemont, Artisan Charcutier, Brittanny, France; and Robert Fagneray, Boucher-Charcutier and former teacher at the Provincial Institute of Food Arts in Namur, Belgium.

Dr Jacques BREVERY, PhD, CEC, AAC

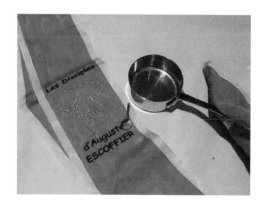

ABOUT THE AUTHOR

Dr. Jacques BREVERY, PhD, CEC, AAC, was born in Belgium in the beautiful region of Ardennes-Lorraine, just a few miles from the scene of the Battle of the Bulges, in 1944. He began his culinary journey at his family restaurant with his mother at the age of 5. He graduated from the famous International Culinary Institute of Libramont-Chevigny, Belgium, in 1975, where he earned a Master Degree with honors and pursued a PhD in Natural Food Science and Anthropology. He is the fifth chef in the family.

He began his hotel journey with Intercontinental Hotel in Brussels and Luxemburg. After 2 years, he was offered the position of Executive Chef (1980) at the Le Floreal Resort in La-Roche-en-Ardennes, in South Belgium. He was there for 5 years, until the hotel closed for major renovations, for 2 years.

In 1981, he won 2nd place in the Prosper Montagne Challenge and was inducted into Les Toques Blanches International by Chef Anton Mossiman in Brussels, 1 year later.

In the Fall of 1985, he took his career to North America when he took a job in Edmonton, Canada, where he worked for 2 years before moving to Florida to join the staff at L'auberge de France, in Palm Beach (Relais Chateaux & Mobil 5-Star).

There Jacques worked for Mr. A. Surmain, MCF, until the hotel closed.

He then joined the Breakers Hotel, working with Chef Michael Norton, CEC, AAC and Chef Scott Gilbert, CWC, where he acquired CWC certification after joining the Palm Beach ACF Chapter in 1988.

He began to lecture and demo on his favorite subjects of food: charcuterie, smoking, curing, binding, pickling, and more for everyone from Chefs and epicurians to home cooks who wanted to learn how to safely use the trimmings and organ meats and make them into enjoyable and attractive dishes to serve patrons and friends.

Chef Brevery became a board member of the Apprenticeship Program Committee where he was the onsite examiner and hot food final exam judge. He was also the Salon Chair for the South-Florida ACF Culinary Salon, from 1988 to 1990.

He then moved to South America to learn Spanish, with a contract with Intercontinental Hotels. He was in Caracas Venezuela at Tamanaco, for the first 6 months, while waiting for the General Manager of the Del Lago Resort in Maracaibo, Peter Fr. Alex, to re-open the 450 room resort, which featured 5 restaurants and was in the process of being renovated. After the opening of the hotel, he remained at Del Lago for 2 years.

In 1993, he returned to Miami, where he took over the kitchens at the Marriott-Caterair-Miami-Airport, 361 & 366, where he ran the daily operations producing 20,000 meals a day for 29 airliners, which generated up to $53 million. There he supervised 450 staff members.

Chef Brevery acquired his American Culinary Federation Certified Executive Chef Certification and became the 1st vice president of the Epicurian ACF Chef Miami Chapter for 4 years. He was also the Educational Chair and was involved in Chef & a child.

In 1993, he was asked by ACF-Team USA manager, Chef Keith Keogh, CEC, AAC, to join the ACF USA team, where he worked on the hot food menu for the world cup Expogast-WACS, to be held in Luxemburg. The World Cup was won by the team in November 1994. He remained with the team until 1996.

In 1995 and 96, he coached and judged the student culinary team from Johnson and Wales University, where he taught charcuterie.

Jacques was awarded the Presidential ACF Medaillon # 2, in 1996, by ACF President, Chef Reimund Pitz, CEC, CCE, AAC.

In 1996, he became the Executive Chef for Arvida at the Broken Sounds Country Club in Boca Raton Florida, where he was the head of 5 food outlets and banquet operations which earned $4 million and was operated by a team of 35.

He was awarded Culinarian of the Year, then Chef of the Year '97 of the Broward County ACF Chefs. In 1998, he was sworn in as an AAC Fellow by Chef Bert Cutino, CEC, AAC, Chairman of **Academy of Chefs**, in the ACF Anaheim Convention.

Every year, Chef Brevery mentored 1 or 2 ACF apprentices by following ACF guidelines. He also gave countless lectures and seminars at culinary schools and technical colleges throughout the United States.

In 1997, he became the Executive Chef at Burt & Jack's Steakhouse in Port Everglades, Fl, working with well known Chef Kevin Hyotte, then named General Operating Manager where he remained until 2004, when it was closed.

They received numerous awards as the top 10 steakhouses in the country, great reviews on the Zagat Guide, Di Rhona, 5 stars from the American Academy of Hospitality Sciences, and medaillon 2000 "Taste of USA".

Throughout the years, he has participated in myriad charity cook-offs, food festivals, and charity events benefiting S.O.S, child hunger, St. Jude's Hospital, Chef & the Child, Broward Navy, Women in Distress and more.

After the closing of the steakhouse, he took a retreat in Western North Carolina in the Smoky Mountains, where he opened a 2000-acre resort for Centex in Tuckasegee (near Cherokee). He worked on the food operations, staffing the kitchen and training, which led to the grand opening event on July 4, 2005, where they cooked for 500 guests in a trailer.

For the past few years, he has been working with his own consulting company helping restaurants in distress, openings, revamping menus, training and designing. He continues to work with culinary schools by giving lectures and is still involved with the ACF. Lastly, he has been taking time to finish writing his book.

In 2011, he was asked to opened two chef societies in Florida:

- **"l'Ordre des Canardiers"** of Rouen–Normandy, France, which promotes food & gastronomie from Normandy and defends the recipe of the "Duck a la rouennaise", which uses a duck press to exhude the blood-juice from the duck carcass. (He was inducted in France on November, 2009.)
- Then the one most prestigious **"Les Disciples d'Auguste Escoffier International USA"** where he was also inducted as chairman for The Florida Society by the Ambassador USA, the honorable Chef Michel Bouit, CEC, AAC, at the first dinner d'automne in Miami at Le Cordon Bleu College of Culinary Arts, last October 19th 2011.
- Finely, February 10th, he was inducted into La "Chaine des Rotisseurs", Baillage of Orlando, and chosen by Chef Dr. Reimund Pitz, conseiller culinaire and Head judge for the Southeast USA, to join him also as a judge for the Skills USA.

CHAPTER 1

Charcuterie

Charcuterie is the art of transforming pork into a variety of different preparations. This array of dishes has traditionally held a very important place in gastronomy, particularly in Europe.

The tradition of charcuterie goes back as far as Greek and Roman times (B.C.). The Greeks and Romans used other meats, such as dog and monkey, in addition to pork, and they used hog bowel as wrappers for their products.

For a long period of history after the Greeks, charcuterie was mainly prepared from pork. Today other meats such as horse, rabbit, game, duck, goose and other poultry, and even fish and vegetable proteins are used in charcuterie, which is still evolving under the influence of modern cuisine.

Consumption of charcuterie continues to grow. Charcuterie provides certain advantages for urbanized populations such as ease of preparation and enhanced shelf life, and can be consumed hot or cold. Certain physical characteristics that result from traditional preservation techniques improve flavor and add nutritional value, and the wide variety of charcuterie products available adds to its gourmet appeal.

According to Christian tradition, every profession has a "patron saint." Saint Antoine (Figure 1.1) is the patron saint of charcutiers. He was born around 250 A.D., and he lived much of his life as a hermit in

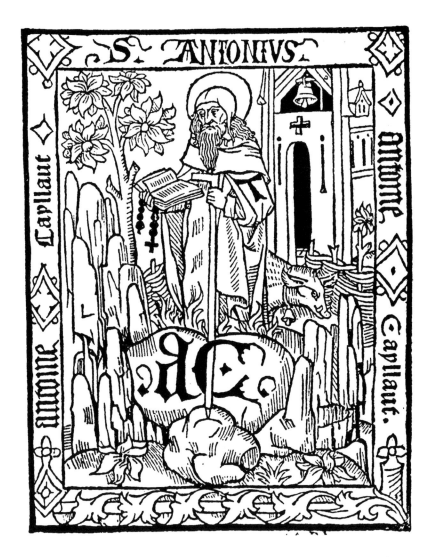

FIGURE 1.1   ST. ANTOINE. (COURTESY OF J.C. FRENTZ —ENCYCL.CHARCUTERY.)

the Egyptian desert, practicing an extremely monastic and disciplined lifestyle. The Archbishop of Alexandria wrote about the life of Saint Antoine in "La Vita d'Athanase" in 357.

Charcuterie is an old and established professional trade in France. The charcutier's trade was difficult to regulate, and it wasn't until the

10th century that strict controls were imposed. In the 15th century, charcutiers were allowed to sell raw pork meats as well as cooked and prepared meats.

The first formal statute of Robert d'Estouville, knight of Guillaume the Conqueror, created the "brotherhood" (*confrérie*) of Master Charcutiers in Paris on January 17, 1475. The confrérie adopted rules for sanitation (*salubrite*), as well as methods for detecting fraudulent sales or products.

The word charcutier comes from "*charcuitier*," or *chaire cuite*—cooked flesh.

On July 18, 1513 King Louis VII decreed that charcutiers could buy pork meats in the markets of Paris and around the kingdom, and could slaughter pigs in order to prepare food products and resell them. The chacutiers officially became "master" tradesmen, turning raw meats into delicacies (sausages, patés, tureens, and galantines) in their kitchens.

The art of transforming pork meat (as well as other meats such as chicken, duck, goose, game, rabbit, fish, beef, veal, and even sheep or horse) varies according to geographical region and knowledge of the artisan. The training (by apprenticeship) is long, and candidates must be apt students as well as physically strong.

This guide does not attempt to exhaustively or precisely address every aspect of the art of charcuterie, but my goal is to share the basics as well as some selected recipes and illustrations. I hope to tantalize the reader into trying some of these *cochonnailles*, or delicacies.

Charcuterie is an art. It demands serious attention on the part of the Chef in order to achieve the necessary aesthetics. Success depends as much on presentation (color, texture, taste, symmetry) and decoration as it does on good execution and technique.

Artisanal charcuterie is constantly evolving, making the best use of modern products and adapting to consumer preferences. This collection of information and recipes will serve as a foundation for perfecting the skills and learning the art of charcuterie.

BON APPETIT!

FIGURE 1.2   PIGS SLAUGHTER MEN CHICAGO 1870.
(COURTESY FELIX REGAMAY-BIBLIOTHEQUE DES ARTS DECO.)

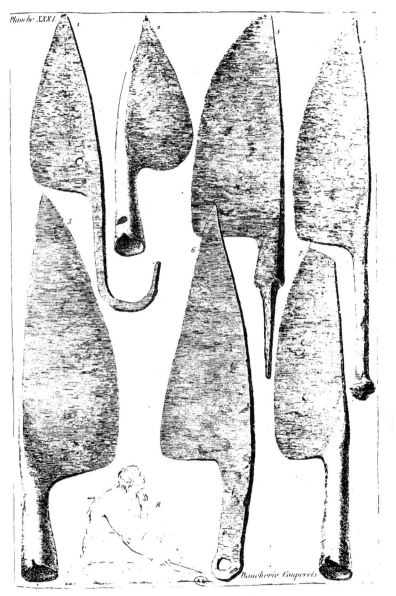

Couperets de boucher de l'Antiquité

Bibliothèque des Arts décoratifs

FIGURE 1.3 KNIVES-CLEAVERS OF BUTCHERS-CHARCUTIERS.
(COURTESY OF BIBLIOTH. OF ARTS DECO—A.L.)

FIGURE 1.4 SIGN OF CHARCUTERIE-SHOP IN ALSACE-FRANCE.

FIGURE 1.5   TONGUE TUREEN, HEADCHEESE, LIVER PATÉS, HAMS-JAMBONEAU PANES, BACON, …

FIGURE 1.6   MORE PATÉS SHOP IN FRANCE, WITH DELICACIES...

FIGURE 1.7   CHARCUTERIE SHOP ALL AROUND EUROPE, EASY FOOD READY TO CONSUME...

FIGURE 1.8 ANOTHER ONE WITH SALAMI AND HAM SERRANO.

CHAPTER 2

Pork

As early as 7000 BC, when nomadic populations in the Middle East began to settle in one place and farm, wild boar were domesticated into the earliest versions of the farm pigs we know today. The adaptable nature and omnivorous diet of this creature allowed early humans to domesticate it much earlier than many other forms of livestock such as cattle. The domestication of pigs occurred even before any known evidence of writing. Pork was consumed in Palestine as early as 1600 BC.

These ancient domestic pigs lived communally with humans, like a guest sharing food with his host. Ancient authors wrote about the voracious appetite of these unusual beasts.

Pigs were mostly used for food, but people also used their hides to make shields and shoes, their bones for tools and weapons, and their bristles for brushes.

Pigs had other roles besides providing meat: their feeding behavior made them useful for searching for roots, and their omnivorous nature enables them to eat human trash, which helped keep settlements cleaner. Even today, people benefit from their sensitive noses that can find the elusive truffle, an underground fungus highly valued for its culinary properties.

Pigs were considered impure in the Old Testament and in the Koran: "You will consider the pig impure because his hoof is split in two; he doesn't graze." (Lev. II, 7). The Bible forbade the consumption of pork meats because of a general preoccupation with purity.

The image of the pig has many representations throughout history. Though considered impure by some, other cultures believed that sacrificing pigs was an act of purification over the powers of evil, since pigs were willing to consume anything.

Today, pork is the general culinary term for the meat products obtained from domestic pigs, which are consumed in many countries. (Figures 2.1 and 2.2) Pork is one of the most commonly consumed meats worldwide, and represents 38% of world meat production. (Figure 2.4). All parts of the pig are consumed, from head to tail (Figures 2.5 and 2.7). Pork is normally cooked and can be processed in several forms. Some of these processing methods are traditional techniques, designed to extend the shelf life of the meat. Curing ham is a way of preserving the meat for later consumption. Cured hams can be cooked and aged like Italian prosciutto or Spanish Serrano Pata Negra, or smoked like Virginia country ham, Belgian Ardennes ham, or the famous acorn hams from Westphalia, Germany. Bacon, pancetta, and some sausages are also forms of cured pork. In Europe and North America before the Industrial revolution, pork was typically consumed in the fall, and oven served at Christmas.

Charcuterie is the artisanal craft of preserving meat, with methods that were developed before the advent of refrigeration. Charcuterie encompasses a broad group of products, such as bacon, ham, sausage, terrines, galantines, patés and confits. Though these preservation methods are perhaps no longer essential, thanks to modern technology, they are still highly valued for the distinctive flavors and textures they produce.

Pork is a nutritious meat, and when well trimmed is also leaner than most domesticated animal meat. Pork is high in thiamine (vitamin B1), and has more myoglobin than chicken (and less than beef). In 1997,

FIGURE 2.1 PIGS THAT HAVE BEEN SPLIT OPEN AND CLEANED AFTER SLAUGHTER (2009).

the National Pork Board started an advertising campaign in the United States, promoting pork as the "other white meat" (chicken and turkey are widely perceived to be healthier than red meat). The campaign was successful and resulted in 87% of consumers identifying pork with the slogan. The slogan was retired in 2011.

Pork Consumption by Country, 2006

	Metric Tons	Pounds Per Capita
China	52.5	88.0
Europe	20.1	96.58
USA	9.0	63.8
USSR	2.6	39.82
Japan	2.5	43.5
Others	12.2	N/A
TOTAL	98.9	N/A

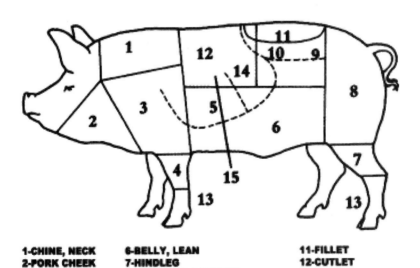

1-CHINE, NECK
2-PORK CHEEK
3-SHOULDER
4-FORELEG
5-FORERIB

6-BELLY, LEAN
7-HINDLEG
8-HAM, HIND QUARTER
9-LOIN
10-PORK CUTLET WITH FILLET

11-FILLET
12-CUTLET
13-FOOT
14-MIDDLE PORK RIB
15-LARD, PORK FAT

FIGURE 2.2   U.S. PORK CUTS.

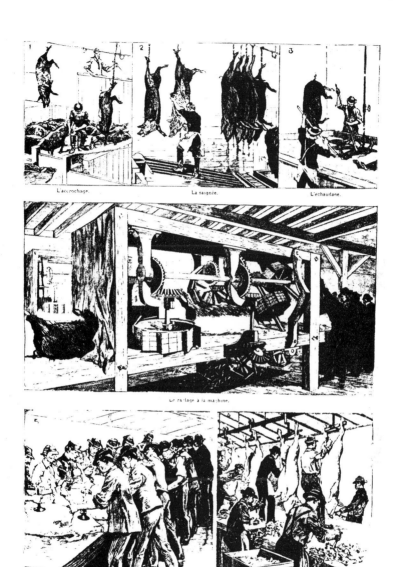

Abattage à Chicago. Gravure de l'« Illustration » (1891).

FIGURE 2.3   CHICAGO SLAUGHTERHOUSE.
(ILLUSTRATION COURTESY OF "L'ILLUSTRATION" 1891.)

Musee Carnavale

Halle d'abattage des porcs vers 1870 aux abattoirs de la Villette

FIGURE 2.4   HALL OF PIGS SLAUGHTER PARIS.
("LA VILLETTE". 1870. COURTESY CARNAVALET MUSEUM.)

FIGURE 2.5   PORK FOR SALE IN A PARIS-RUNGIS MARKET (2009).

Le porc

FIGURE 2.6   PORK AS COMMUNAL FOOD.
(COURTESY OF "CHARCUTERY ENCYCLOPEDIA, J.C. FRENTZ.)

FIGURE 2.7   FROM HEAD TO TAIL…

CHAPTER 3

Nutrition Tables

COMPONENTS IN VARIOUS MEAT GRADES

TABLE 3.1

The nutrition content of different kinds of meat depends greatly upon the type of animal and where it was raised. Meat is graded according to its quality, with ratings such as EAA (prime), IA (choice), IIA (select), etc.

Class	EAA	IA	IIA	IB	IIIA	IIB	IC	IIIB	IIC	IV
Bone (%)	12.9	13.7	12.6	14.2	12.0	13.3	14.6	12.7	13.1	12.6
Spine (%)	5.8	6.3	5.9	6.4	5.9	6.4	6.6	6.2	6.2	5.8
Rest (%)	2.6	2.8	2.4	2.7	2.7	2.7	2.8	2.4	2.6	2.7
Muscle/Fat (ratio)	2.80	2.13	1.55	1.83	1.34	1.49	1.63	1.22	1.23	1.09
Muscle/Bone (ratio)	4.49	3.80	3.77	3.46	3.78	3.47	3.22	3.43	3.28	3.26

TABLE 3.2
Nutritional Composition of Pork and Beef

	% Water	% Fat	% Protein	% Collagen	Total% Soluble Carbohydrate
Pork					
Lean, without visible fat and nerves	74	5	20	2	0
Lean, with nerves, without visible fat	69	11	19	5	0
Lean, with ±30% visible fat (no skin)	52	32	14	12	0
Jaw	67	16	16	5	0
Shank	71	9	20	4	0
Fat back	9	88	2.5	2	0
Bacon	16	78	6	2	0
Throat	30	60	7	2.5	0
Panne	8	90	1.5	1	0

TABLE 3.2 (continued)
Nutritional Composition of Pork and Beef

	% Water	% Fat	% Protein	% Collagen	Total% Soluble Carbohydrate
Ratis	28	67	5	3	0
Liver	71	4	21	3	3
Heart	70	8	17	5	0
Tongue	80	3	15	8	0
Whole blood	91	0	8	0	0.5
Skin:					
• Raw	44	16	40	35	0
• Cooked: 15 min.	51	11	33	39	0
• Cooked: 30 min.	58	6	28	33	0
• Cooked: 60 min.	64	6	21	28	0
• Cooked: 90 min.	69	55	20	25	0

continued

TABLE 3.2 (continued)
Nutritional Composition of Pork and Beef

	% Water	% Fat	% Protein	% Collagen	Total% Soluble Carbohydrate
Beef					
Lean, without visible fat and nerves	75	2	22	2	0
Jaw	70	12	18	10	0
Fat	10	84	4	3	0
Liver	70	4	21	4	4
Heart	68	10	15	4	0
Tongue	64	18	17	5	0
Blood	91	0	8	0	0.5

TABLE 3.3
Average Nutritional Composition of Other Meats and Ingredients

	% Water	% Fat	% Protein	% Collagen	Total% Soluble Carbohydrate
Variety Meats					
Horse (very lean)	76	1	22	2	0
Turkey					
• White meat, cleaned	74	2	25	0.5	0
• Dark meat, cleaned	73	5	21	1.5	0
• Skin	39	50	11	6.5	0
Meats separated by machine	68	13	17	1.4	0
Rabbit	74	2	22	1.5	0
Hen	70	7.5	21.5	2.2	0
Chicken					
• White meat, cleaned	74	2	23.5	0.5	0
• Red meat, cleaned	72.5	8.5	19	1.5	0

continued

TABLE 3.3 (continued)
Average Nutritional Composition of Other Meats and Ingredients

	% Water	% Fat	% Protein	% Collagen	Total% Soluble Carbohydrate
Skin	37.5	47	10.5	6	0
Mechanically separated meat	64	21.5	14.5	2	0
Cooked, mechanically separated meat	74.5	6.5	18.5	2	0
Veal					
Leg	68	12	19		0
Shoulder	70	10	19		0
Breast	62	20	16.5	4.5	0
Other Ingredients					
Amidon, starch	10	0	0	0	90
Caseinate	5	3	90	0	1
Vegetable proteins	5	0	70	0	24
Water	100	0	0	0	0
Wheat flour	14	0	11	0	70

TABLE 3.3 (continued)
Average Nutritional Composition of Other Meats and Ingredients

	% Water	% Fat	% Protein	% Collagen	Total% Soluble Carbohydrate
Gelatin Powder	10	0	88	88	0
Isolated vegetable proteins	4	0	93	0	3
Lactoserum powder	3	0	13	0	72
Milk					
• Fresh	90	4	3.5	0	5
• Milk powder	5	0	35	0	52
Levure alimentaire	5	7	50	0	30
Eggs					
• Whole	74	12	12	0	1
• Egg white	87	0	11	0	1
• Yolk	49	32	17	0	1

continued

TABLE 3.3 (continued)
Average Nutritional Composition of Other Meats and Ingredients

	% Water	% Fat	% Protein	% Collagen	Total% Soluble Carbohydrate
Powdered Eggs					
• Whole	7	42	6	0	3
• Egg white	3	0	31	0	6
• Yolk	3	61	86	0	2.5
Blood Protein					
• Fresh plasma	91.5	0	7.0	0	0.1
• Dry plasma	7	0	72.0	0	1
• Whole plasma	70.0	0	29	0	0
Sugar	0	0	0	0	99

TABLE 3.4
Plant and Animal Fat Sources

Fats	Natural Origin
Butter	Milk
Corn oil	Grains/seeds
Sunflower oil	
Soy	
Pork clarified lard	Pork fat
Beef clarified lard	Beef fat

	Temperature	Packaging	Time of Holding	Observations
		Meats		
Beef				
Beef: carcasses	4	None	10 to 14 days	
Beef: carcasses	−1.5 to 0	None	3 to 5 weeks	
Beef: carcasses (boneless)	−1.5 to 0	None, with 10% CO_2	9 weeks	
Beef: slabs, boneless	−1.5 to 0	Crayovac	12 weeks	
Beef: pieces	4	Plastic wrap	1 to 4 days	
Retail Cuts				
Retail: pieces	4	Crayovac	14 days	
Retails: pieces	2		9 to 12 days	80% O_2+; 20% CO_2
Ground Beef	4	Plastic wrap	24 hours	

continued

	Temperature	Packaging	Time of Holding	Observations
Ground Beef	4	Crayovac	7 to 14 days	
Ground Beef	2		3 to 5 days	80% O_2+;j 20% CO
Pork				
Carcass	4	None	8 days	
Carcass	−1.5 to 0	None	3 weeks	
Slabs	−1.5 to 0	Crayovac	3 weeks	
Retail Pieces	4	Plastic wrap	3 days	
Ground	4	Plastic wrap	24 hours	
Slices of Charcuterie Products	4		3 to 6 weeks	Depending on water retention and bacterial quality
Wiltshire Bacon	4	None	3 to 5 weeks	

	Temperature	Packaging	Time of Holding	Observations
Lamb				
Lamb	4	None	1 to 2 weeks	
Lamb	−1.5 to 0	None	3 to 4 weeks	
Sheep	−1.5 to 0	Crayovac	10 weeks	
Veal	4	None	6 to 8 days	
Veal	−1.5 to 0	None	3 weeks	
Organs	−1.5 to 0	None	7 days	
Fowl				
Fowl	4	Plastic wrap	7 days	
Fowl	−1 to 0	Plastic wrap	2 weeks	
Fowl	−2.2 to −1	Plastic wrap	3 weeks	
Fowl	−2	Plastic wrap	3 to 4 weeks	

Note: Humidity of 85 to 95% needed normally unless usage of special packing like Crayovac or other.

CHAPTER 4

Salt

Salt is an essential component of our diet. The human body needs at least one gram of salt per day to function properly, though excess salt intake can contribute to health problems like high blood pressure. Nowadays gourmet salts are available from around the world.

Salt is also fundamental in the processing of charcuterie. Salt has been used for centuries to preserve, color, and season meats. Salt has three main purposes in charcuterie. It improves the flavor of the meat, it helps the meat retain water, and it helps to preserve the meat by inhibiting the growth of bacteria. But there are some limitations—the amount of salt needed to completely prevent bacterial growth would ruin the taste of the meat, and salt produces an undesirable grey color. To overcome these issues, nitrates and sugar are added for a more attractive pink/red color, as well as to improve protection against bacteria, and to provide characteristic flavor. Nitrites also delay the development of botulinal toxin as well as the development of rancid flavors during storage. Meat that is preserved in this way is known as "salt cured" meat.

By law, the amount of sodium nitrite that can be used for curing is limited to 0.6%. The limit for sodium chloride is 99.4%.

FIGURE 4.1　ALL MISE EN PLACE FOR CURING A SMALL PIECE AT HOME.

BRINE CURE

I was taught in school that the use of saltpeter (sodium nitrate) for preserving meat dates back to the Middle Ages, or even earlier. Saltpeter occurs naturally and has long been harvested from stone outcroppings and walls as well as from desert salts.

In 1891, a scientist named Polenske discovered that it was the nitrites in saltpeter that caused the color change in meat. In the 20th century, scientists further clarified the mechanism by which this occurs—nitric oxide binds to the iron in the oxygen-carrying myoglobin cells, which fixes them with a red color. Saltpeter is still used for large pieces of meat like sausages and hams because its effect on color and appearance lasts longer than that of sodium nitrite, a more modern additive. Sodium nitrite is a white (sometimes yellowish) crystalline powder that

FIGURE 4.2　BRINE CURE WITH A WEIGHT. (COURTESY OF J.C. FRENTZ.)

dissolves well in water. A good recipe for smaller pieces of meat is a mixture of 0.6% sodium nitrite and 2 g of sugar per pound.

Brine curing is a method of meat preservation that involves immersing meat in a prepared mixture of water, salt and spices. Brine solutions can also be injected directly into the meat.

Sugar plays an important role in the brine cure process because it serves as a nutrient for the bacteria, which transfer the saltpeter into nitrite, which reacts with the meat to form nitric oxide, which helps to fix the natural color of the meat. Sugar also balances the acrid taste of the saltpeter and/or nitrites.

Regular sugar mixed with saltpeter were the traditional ingredients used for brine curing, while reduced sugar (dextrose) mixed with sodium nitrite is more often used for brining today (with faster results). Regular salt gives preserved meat a grey or dark color, while sodium nitrite results in a pink/red color.

Sodium nitrite, sodium or potassium nitrate (saltpeter), and sugar are ingredients to use **with caution**. The dosages for their use in charcuterie are precisely regulated. It is important to adhere to the dosages in these recipes, which are in compliance with the international regulations for charcuterie processing. Consulting local health codes and regulations and the food safety codes for particular ingredients is advisable.

BRINE CURE RECIPES

(See Table 4.1)

The following recipes are for 10 L (2.67 gal.) of water at 14 degrees Baumé/1107.

Use only pure, non-chlorinated water, and about 210 g salt/quart or L of water.

Brines can be prepared with heat in order to help the ingredients dissolve. Boil the water with all of the ingredients, except the sugar, until they are dissolved. Remove from heat, skim any solids from the surface, and add the sugar. Transfer to a clean container, cover (to avoid any contaminants) and let cool. Store mixture until ready to use. Decant and aerate brine mixture if needed.

If using the cold brine method (no heat), make sure that all of the ingredients are completely dissolved by mixing thoroughly.

It is recommended that very fresh pork meats (spine, breast, round) be immersed for 48 hours for a better coloration.

TABLE 4.1

I.	
Kosher salt	650 g or 22.9 oz.
Sodium nitrite	650 g or 22.9 oz.
Saltpeter	20 g or 0.7 oz.
Sugar	100 g or 3.5 oz.
Aromatics	
Room temperature	(4–6°C, or 40–42°F)
II.	
Sodium nitrite	1300 g or 45.8 oz.
Saltpeter	13 g or 0.46 oz.
Sugar	100 g or 3.5 oz.
Infusion* see below.	
Room temperature	(4–6°C, or 40–42°F)
III. (*Old style recipe*)	
Kosher salt	1300 g or 45.8 oz.
Saltpeter	40 g or 1.4 oz.
Sugar	100 g or 3.5 oz.
Room temperature	(8–12°C or 43–52°F)

* In all cases it is advised to use an infusion, particularly for the cold brine-cure.

BRINE CURE INFUSION

Dosage for 3.5 gal.: ½ gal. aromatics and 3 gal. for dissolving salts, saltpeter.

Aromatics	
Fresh Thyme	½ oz.
Dry Thyme	¼ oz.
Fresh bay leaves	⅛ oz.
Dry bay leaves	⅛ oz.
Fresh unpeeled garlic (germ removed) and diced	1½ oz.
Juniper berries	⅛ oz.
Whole pepper	¼ oz.
Cloves	⅛ oz.

Place the aromatics in a half-gallon of water and boil for 10 minutes, covered. Dissolve salt, saltpeter and sugar, or sodium nitrite and dextrose in the other 3 gal. of cold water. Strain the aromatic infusion and add to the 3-gal. mixture. Decant and aerate if necessary. Check the density of the saumure (brine) with a Baume densitometer. Take care to maintain cleanliness with all of the equipment and ingredients used.

SPICE/SALT MIXTURES

This is an example of a spiced salt that I use for my foie gras:

12 to 15 g per kg (2.25 lb.) of liver:

Nitrite salt	10 g	10 g	10 g	10 g
White pepper	1.5	1.5	1.5	1.5
Nutmeg	0.1	—	—	—
Mace	—	0.1	0.1	0.1
Ground cloves	—	0.02	0.02	0.02
Thyme	0.2	0.05	0.05	0.05
Bay leaves	—	0.05	0.05	0.05
Marjoram	—	0.05	0.05	0.05
Cardamon	0.1	0.05	—	0.1
Basil	—	0.02	—	—
Coriander	—	0.02	—	—

Note: Quatre epices (for preparations such as Duck "a la Rouennaise"): Cinnamon: 35%, nutmeg 15%, allspice 15%, pepper 35%.

Nowadays the mixes finds in the markets are mostly as:

Allspice ground: 14%
Nutmeg: 10%
Mace: 10%
Cinnamon: 32%
Cloves: 18%
Ginger or caraway: 16%
Indian flavors (for spice bag): cardamom, star anise, turmeric, saffron, cloves, and black pepper
Savory Charcuterie flavors: whole savory, fresh rosemary, fresh bay leaves, thyme, marjoram, and wild garlic
Health Mix: fresh mint, cayenne pepper, thyme, lemon (quarter squeeze), garlic, ginger

FLEUR DE SEL

The name *fleur de sel* translates to "flower of salt" in French. This expensive salt is harvested from several coastal areas around the world, in places such as the pristine salt marshes of Brittany and from the Arabian Sea at the foothills of the Himalayas. Fleur de sel is widely available throughout France, costing only about $3 per kilo at French supermarkets. This unique sea salt is lighter in color than grey sea salt, and the crystals are finer in texture. Though unrefined, it is very pure and retains many of the minerals found in seawater. Some call fleur de sel the "diamond of the sea" or "white gold."

Fleur de sel is a gift from nature and is valued for its many reputed healing properties, including restoring calm and balance in those who ingest it. It has very pure crystals, which contains a natural balance of life-enhancing minerals. Fleur de sel is harvested from the surface of crystallizing ponds by hand, preserving the harmonious ecological partnership between humans the environment. The salt crystals of fleur de sel are then left to dry naturally in the sun.

SMOKED SALT

Smoked salt is a product made by combining traditional wood flavor with sea salt to create a unique spice that goes well in many recipes. Natural smoked salt is slow smoked over real wood, giving it an authentic clean taste. No artificial ingredients like liquid smoke or artificial flavors are added. Smoking salt takes time, just like smoking meat. The salt absorbs flavor as the smoke resins coat the salt crystals. It takes between72 and 95 hours to achieve the best results, and this unusual salt is only produced in a few countries. Smoked salt can be use alone or with other spice blends. It has been described as tasting "like a bonfire."

TABLE 4.2
Desirable and Undesirable Brine Characteristics

Characteristics	Ideal	Undesirable
Color	Yellow-orange to brown-red	Green reflects brick-red color
Turbid	Medium or none	Very strong
Form fleecy (soft, white)	None or light	Very dense
Foam or scum on surface	None	Present
Odor	None or meat or stock odor	Musty, acrid, putrid or abnormal odors.
Taste	Typical, aromatic	Acrid, insipid, putrid
Salt concentration	13–22 degrees Baumé (1198/1179)	Greater than 22°Bé
Nitrites	600–1200 mg/L or 0.02–0.04 oz./ 35.5 liq oz.	Lower tham100 mg/L or 0.0033 oz./ 35.5 liq oz.; upper to 6.2 oz.
Ph	5.6–6.2	Lower than 3.0 and greater than 6.0 for slow brine cure. Lower than 4 and greater than 10 for fast brine cure.
Temperature	5–7°C; 40–42°F	

CHAPTER 5

Binding Agents and Fillers Used in Charcuterie

HYDROCOLLOIDS

Agar-Agar: Agar-agar is a gelatinous plant material that is also known as "Chinese Gelatin" or "Japanese Isinglass." Agar-agar is made of polysaccharides obtained from red seaweed found in coastal Japan, South America, Portugal and Spain. It is used as a culture medium, as a thickening agent in food, and as a vegetarian substitute for gelatin. Agar is a gel at room temperature, melting at approximately 70°C (160°F), and is capable of absorbing 300–500 times its weight of water.

Alginate (E401-E404): Sodium alginate is the salt of alginate, a polysaccharide found in the cell walls of brown algae (which are harvested off the coast of Brittany, among other places). It is used to produce a stable, translucent, and flavorless gel in meats and dairy products.

Carob Powder (E410): Carob powder is made from the seedpod of the *Ceratonia siliqua* tree, or Carob tree, which grows well in the

43

Mediterranean. Carob powder becomes viscous when mixed with hot water, and can form a gel in the presence of xanthan gum. It's used as a food-thickening agent in ice creams and mayonnaise, and in meat processing.

Carrageenan (E407): There are several varieties of carrageenans, all of which are polysaccharides obtained from red seaweed. In Ireland and Scotland, a traditional flan-like dessert has been prepared for hundreds of years by obtaining these carrageenans from the red seaweed that grows along the coast. One variety reacts with dairy proteins like casein to form gels. Carrageenans are used in meat processing and to thicken dairy products.

Gum Arabic (E414): Gum arabic is a natural gum obtained from the acacia tree in Africa and the Middle East. It produces a soft gel when dissolved in water, and is stable in emulsions. The use of gum arabic is not allowed in meat processing, but it is used in candy production and has many industrial uses.

Xanthan Gum: Xanthan gum is produced by bacterial (*Xanthamonas campestris*) fermentation of sugars. It's used as a thickening agent and stabilizer in meat products, dairy products, sauces, and frozen and gelled desserts.

STARCHES

Vegetable starches are commonly used as thickeners, water binders, emulsion stabilizers, and gelling agents in charcuterie. There are many varieties of starches, processed with different techniques, all with specific applications. The choice of a particular starch additive depends on cooking temperature, desired texture, color, and storage and processing conditions. Starches are insoluble at room temperature but absorb water when heated. When chilled, starch additives can form gels. When stored at a cold temperature, the starch can return to its insoluble state and may loose its thickening properties.

Barley or Wheat Flour: These flours have variable properties (65–75% starch and 8–14% gluten) as enzymes hydrolyze the starches.

Unmodified Starches: Potato starch and cornstarch are examples of unmodified starches—they are not treated chemically. They produce an opaque gel that is more sensitive to processing and freezer storage. Thickening properties diminish in acid conditions. These starches are used in charcuterie in sauces that will be consumed immediately. Potato starch has better flavor and has superior thickening properties. Cornstarch is mostly amylopectin and does not form a gel, but works as a thickening agent in solution and is stable as well as translucent.

Modified Starches: Modified starches are starches that have been chemically treated to make them more stable at high temperatures (such as those used in pasteurization) in an acid environment and in homogenization.

Maltodextrins: Maltodextrin is a product made from the partial hydrolysis of starch. Maltodextrin has different strengths and dissolves in a small amount of liquid, is a good thickener, and is stable without being too viscous.

Pregelatinized Starches: These starches are produced from different vegetable sources. They are precooked and thicken or gel without the application of heat, so that they can be used in cold food preparations.

ANIMAL PROTEIN BINDERS

Casein: Casein is a milk protein that has gelling properties, is essential to the coagulation of milk in cheese formation, and can retain water.

Egg Whites: Egg whites are the albumen component of eggs. They are mostly water and contain 10% protein. Egg white coagulates at 150°F and can be used to form gels.

Plasma: Plasma is a blood fluid that can be used as an emulsifier or gelling agent. It also contains albumen and coagulates at 150°F.

Collagen/Gelatin: Gelatin is a protein substance obtained from the hydrolysis of collagen, which breaks down at high temperatures. Collagen is used as a gelling agent and produces a firm and long-lasting gel. High levels of salt can prevent gel formation.

VEGETABLE PROTEIN BINDERS

Gluten: Gluten is a protein found in wheat and other grains. It is used as a binder in cooked meat preparations and coagulates at 190°F. Gluten is also used to add elasticity to bread products.

Maltose: Maltose is a disaccharide (sugar) that is produced when amylase breaks down starch. It's used in food products as a sweetener and is also known as malt sugar.

Pectin: Pectin is a polysaccharide found in the walls of plants. It's normally obtained from fruits (oranges, lemons, apples) and used as a gelling agent, especially in jams and confections, ice creams, and to stabilize sauces and emulsions. Pectin is not permitted for use in meat processing.

Soy: Soy flour is a protein powder that is produced from soybeans. Soy protein is often used as an "extender" in meat products, as a low cost source of protein. Soy coagulates at 150°F.

TABLE 5.1
Thickeners, Gelling Agents, PolySaccharides

Name	Origins	Made In	Chemical Composition	Usage
Agar–Agar	Red seaweed	USA, Japan	Agarose, Agaropectin	Confection Microbiology
Carrageenan	Red seaweed	USA, Japan, Asia	Galactose polysaccharises	Meat products Gelified milk
Alginate	Brown seaweed	USA, UK	Polysaccharide Guluronic acid Mannuronic acid Galacturonic acid Rhamnose	Meat products Gelified milks Candies Preserves Ice cream Sauces
Gum Arabic	Acacia tree	Africa, Middle East	Saccharides Glycoproteins	Confection Printmaking

continued

TABLE 5.1 (continued)
Thickeners, Gelling Agents, PolySaccharides

Name	Origins	Made In	Chemical Composition	Usage
Pectin	Fruits: apples, oranges, bananas	All temperate regions	Polysaccharides	Candy Preserves Ice cream Sauces
Carob flour	Carob tree	Mediterranean	Legume	Meat products Gelified milk Mayonnaise Sauces

Gaur gum	guar beans	India, USA	Polysaccharide (galactomannan)	Meat binder Gelified milks Ice cream Mayonnaise
Xanthan gum	Bacteria—*Xanthamonas campestris*	Industrial production	Polysaccharide	Many food applications
Starches	Tubers, grains	Worldwide	Glucose	Charcuterie confection
Cellulose gum	Plant trimmings Cotton	Industrial	Cellulose derivative	Processed meats

TABLE 5.2
Binders

Types	Origins	Effect
Starches	• Potato • Manioc • Corn • Wheat flour • Modified starch • Maltodextrin	Thickening agent Gelifying agent
Hydrocolloids	• Gaur • Carob • Alginate • Carrageenans • Xanthan gum	Thickening agent Gelling agent Colloidal protection effect
Other Thickeners Used in Meat Processing		
Egg yolk	Chicken eggs	Emulsifier
Mono and di-glycerides	Animal and vegetable fats	Emulsifier

CHAPTER 6

Master Pastry Recipes for Charcuterie

BASIC PIE PASTRY

INGREDIENTS

1 lb. (454 g) flour
0.2 lb. butter (90 g or 6.5 T) butter, cubed and softened
0.5 oz. (14 g) kosher salt
2 medium eggs at room temperature
5 oz. (142 g or ⅝ C.) cold water

PREPARATION

Place the flour in a bowl or directly on the counter and make a well in the center.
Add the pieces of butter and salt to the well.
Stir in half of the water and the eggs.

Work the liquid ingredients into the flour with your fingers, adding the rest of the water gradually, until you obtain a soft, smooth dough.

Wrap tightly with plastic wrap and keep refrigerated until ready to use.

FOR PATÉ IN A PASTRY CRUST

Follow recipe for pie pastry, using only 3 oz. (85 g, ⅜ C.) of water.

Optional: use hot water (50°C or 125°F) instead of cold water to give the dough a smoother (less flaky) consistency, making it easier to slice.

FOR CRUSTADES (PATÉ SABLÉE)

Follow the recipe for basic pie pastry, but start by sifting the flour over the cubes of cold butter. Quickly work flour into the butter with fingers until mixture is crumbly.

Add 6 oz. of cold water and eggs into the flour/butter mixture. Working quickly, use your hands to fold the mixture over itself to incorporate air into the dough.

Refrigerate dough until ready to use.

PATÉ FEUILLETÉE (PUFF PASTRY)

The French name of this dough means "leaf-like" because it has lots of thin delicate layers or "leaves". This dough has the same simple ingredients: flour, butter, water, and salt, but the mixing technique is quite different, in order to obtain the flaky layers.

INGREDIENTS

1 lb. (454 g) flour
14 oz. (397 g) unsalted butter
10 oz. (296 mL) water
0.5 oz. (14 g) kosher salt

PREPARATION

CLASSIC METHOD:

Mix the flour with the water and salt.

Flatten the cold butter into one rectangle by rolling it out between two pieces of plastic wrap.

Roll out the dough into a rectangle twice as large and place the butter in the middle of the dough rectangle.

Fold the dough over the butter, completely enclosing it.

Gently roll out the dough/butter package into a rectangle.

Fold dough into thirds as if it was a letter going into an envelope.

Turn the dough 90 degrees and roll out into a rectangle.

Fold the dough into thirds again.

Wrap dough in plastic wrap and chill thoroughly.

Repeat procedure until dough has been folded into thirds 5 times, chilling dough between turns if necessary.

Chill dough until ready to use.

QUICK METHOD:

Cut the cold butter into cubes and mix butter into flour with fingers, before adding the water and salt to the dough.

Dough will be crumbly, but use a pastry scraper to make a couple of rough folds in half.

Once dough is more workable and smooth, give it one quick envelope fold (into thirds), then chill thoroughly.

Roll dough out again and make two more envelope folds, then chill until ready to use.

3-DUTCH METHOD:

Cut the cold butter into cubes and mix the butter into one third of the flour with your fingers until mixture is crumbly.

Mix the remaining ⅔ of the flour with the water and the salt and roll out resulting dough into a rectangle.

Place the crumbled butter/flour mixture in the middle of the rectangle, and fold dough over to enclose it, as in the classic method.

Roll out dough and make envelope folds in the same way.

BRIOCHE DOUGH

Brioche is a fermented (yeast-risen) dough that requires a long preparation time. The dough must rest for 12 hours before use.

INGREDIENTS

1 lb. (454 g) all purpose flour
4 medium eggs
8 oz. (227 g) unsalted butter
3 oz. (3/8 C.) milk
0.5 oz. (14 g) kosher salt
0.35 oz. (10 g) yeast
0.35 oz. (10 g) sugar

PREPARATION

Make the "levain": warm the milk until just hot to the touch and stir in the yeast.

Stir in ¼ of the flour until well mixed, then let mixture rest in a warm place until bubbly.

Mix the remaining flour with the water, salt, sugar, eggs, and butter.

Add the levain and knead until dough is smooth, shiny, and pulls away from the bowl or counter (dough should not be sticky).

Cover the dough and let rise in a warm place until doubled in size, about 2 hours.

Punch down the dough and refrigerate until ready to use.

BRIOCHE DOUGH FOR SAUSAGE (LYON) (BOUDIN)

This dough is lighter than regular brioche and pairs well with whole sausages.

INGREDIENTS

1 lb. (454 g) all purpose flour
11 oz. (312 g) butter, softened
6 eggs
3 oz. (3/8 C.) warm water
3 oz. (85 g) sugar
0.5 oz. (14 g) kosher salt
3 oz. (85 g) yeast

PREPARATION

Make a well in the center of ¼ of the flour.

Add the yeast with the warm water into the well.

Mix dough and shape into a ball.

Cover with a towel and let rise until double in volume.

Using the remaining flour, make a well in the center and add the salt and sugar (dissolved in a little bit of water) and two of the eggs.

Mix the dough by lifting it and beating in order to incorporate as much air as possible.

Add two more eggs and mix well.

Add remaining 2 eggs and mix.

Add the butter gradually, kneading after each addition, and knead until dough becomes very elastic.

Place the dough out on a counter and knead the levain into the dough, kneading until dough is smooth and elastic once more.

Dust the dough with flour and let it sit at room temperature until doubled in volume.

Punch down dough, fold it over itself a few times, wrap tightly with plastic wrap, and refrigerate until ready to use.

CHAPTER 7

Sausages

A BRIEF HISTORY OF SAUSAGES FROM AROUND THE WORLD

The word sausage is derived from the Latin word *salsicia*, which referred to ground pre-salted meat, which in turn was based on the Latin word *salsus,* meaning "salted." Sausage preparation goes back to the time of the Ancient Greeks and Romans. Homer even mentioned sausage in the *Odyssey.* Early cooks stuffed roasted intestines into stomachs. In ancient Italy, Lucania (now Basilicata) was known for its *lucanica* sausage.

Sausage is generally made of ground meat (typically pork and beef) mixed with herbs and seasonings. The *charcutier* profession is the oldest profession (*métier*) in the world. Making sausage (*saussiche,* in French) is a way to use up smaller and less valuable pieces of meat and organ meats.

Nowadays, sausage is a food staple in most cuisines, often enjoyed on the weekend and cooked on the grill. There are many charcuterie shops selling gourmet sausages containing cheese, sundried tomatoes, spinach, dried fruits, nuts, a multitude of spices, and there are even vegetarian and seafood sausages.

Many regional cuisines have their own characteristic sausage, such as Spanish chorizo, Andouille from Le Mans, saucisson sec from the French Alps. Italian fennel sausage, German bratwurst and weisswurst, Greek loukánika, English cumberland sausages (bangers), Mexican salchicha oaxaqueña, American hotdogs, North African merguez, cevapcici from southeastern Europe, Argentinian morcilla, Chinese Lap Cheong, Scottish stornaway black pudding, Bulgarian lukanka salami, Finnish makkara, Hungarian kolbasz and majas hurka, Swiss cervelat, Turkish sosis, longaniza from the Philippines, Korean blood sausage, South African boerewors, and even a fish-based "sausage" called kamaboko from Japan. There are many more. The most renowned sausages are from Italy (with or without fennel) and the Toulouse region in France.

Sausage is typically stuffed into a casing, depending on the variety. Sausage casings are usually pork, beef or lamb intestines, but synthetic ones are also available (made with beef collagen). Making sausage is a form of food preservation, as the sausages can be cured, dried, or smoked. I also like to mix sausage meat with caul fat to form sausage patties and links, which are especially good for breakfast.

My father always said that an efficient chacutier should be able to turn a 220 lb. (100 kilo) slaughtered pig into 250 lb. (115 kilos) of sausages, patés, bacon, headcheese, blood sausage, and hams (adding spices, condiments, and seasonings), which is very practical as well as profitable. Nothing is wasted.

The meat to fat ratio in sausage depends on the variety, but most contain 30% to 50% fat by weight. European styles do not have bread fillings and are 100% meat, fat and seasoning. In England and other countries, sausages may contain up to 25% of filler, such as starches and bread products.

Different countries and regions classify and control sausage production and ingredients according to local traditions and regulations. For example, in most English speaking countries the categories of sausage include cooked sausages, cooked smoked sausages, fresh sausages, fresh smoked sausages, dry sausages, bulk sausages, and vegetarian sausages. Germany has some 1200 different types of sausage, including raw sausage, cooked sausage, and precooked sausage. In Italy there are raw sausages, cured sausages, cooked sausage, blood sausages, liver sausages, salami, and cheese sausages. France has hundreds of types of sausage as well, and so it goes with each country.

Let's get to work making different varieties of sausage using the recipes in the following chapter.

FIGURE 7.1   AN ARRAY OF SALAMIS AND ROPE-CURED SAUSAGES FROM EUROPE.

FIGURE 7.2   TRADITIONAL "BOUQUET GARNI" OF ROSEMARY, BAY LEAF, AND THYME.

RECIPES

FARCE A GRATIN—(PASTE ALSO USED FOR GALANTINES OR FANCY PATÉS)

Chef's note: This mixture is used to improve the flavor and texture of certain charcuterie products. It is finely chopped and used in galantines, ballotines, patés en croûte, and mosaic tureens.

INGREDIENTS

4 oz. goose or duck fat

10 oz. pork breast, cut into cubes

12 oz. fresh chicken liver, chopped

3 oz. minced fresh white mushrooms

1 oz. finely chopped purple shallots

4 oz. foie gras, cubed

4 oz. Cognac

4 egg yolks

¾ oz. salt

0.2 oz. ground grey pepper

0.2 oz. mixed fine spices, see "epices-marie" (blood sausage, recipes)

PREPARATION

Process all meats in meat grinder, using the medium plate.

Mix together well.

Use a food processor to chop meat into a fine paste.

Mix in seasonings, eggs, and cognac.

Keep mixture refrigerated until use.

FIGURE 7.3   MEAT GRINDING FOR SAUSAGES.

GROUND SAUSAGE BASE MEAT MIXTURE

INGREDIENTS

3 lb. lean pork shoulder (Boston butt, picnic shoulder)
1 lb. cold fatback
or
2 lb. lean pork, and
2 lb. flank
2 oz. ice water
1.5 oz. kosher salt
0.2 oz. white or grey ground pepper

FIGURE 7.4   HOME MEAT GRINDER.

PREPARATION

Grind all of the meats and the cold fatback with a meat grinder, using
a disk with large holes.

Mix well.

Store in refrigerator until needed.

TABLE 7.1
Sausage—Weight Filling in Grams for Sausages Contents by Size and Diameter for Casings

cm	Ø 43	47	50	55	60	65	70	75	80	85	90	100	105	120
15	120	140	165	190	—	—	—	—	—	—	—	—	—	—
17	160	180	200	240	250	275	—	—	—	—	—	—	—	—
19	190	210	240	290	300	350	—	—	—	—	—	—	—	—
21	225	250	280	340	365	425	460	520	560	—	—	—	—	—
23	260	290	320	390	430	500	540	610	650	675	—	—	—	—
25	290	330	360	440	495	575	610	700	750	800	870	980	1100	—
27	320	370	400	490	560	650	690	790	850	925	1025	1165	1275	1425
29	350	410	440	540	625	725	770	875	950	1050	1150	1300	1450	1650
31	380	440	480	590	690	800	860	975	1050	1175	1275	1475	1625	1875
33	410	480	520	640	755	875	950	1060	1150	1300	1425	1640	1825	2100
35	440	520	560	690	820	950	1030	1150	1250	1425	1550	1780	1990	2310
37	470	560	600	740	885	1025	1110	1240	1350	1550	1675	1950	2175	2575

Ø 38	43	47	52	58
160	190	210	250	320 g
195	240	260	310	400 g
225	275	310	370	480 g
265	315	360	440	560 g
300	355	410	510	640 g
335	390	460	570	720 g
370	430	510	630	800 g
405	470	560	690	880 g
440	510	610	750	960 g
475	550	680	810	1040 g
505	590	710	870	1120 g
540	630	760	930	1200 g

Ø																		
39	500	600	640	790	950	1100	1200	1330	1450	1675	1825	2125	2350	2800	575	670	910	990	1280 g
41	530	640	680	840	1015	1175	1275	1420	1550	1800	1950	2275	2550	3025	605	710	860	1050	1360 g
43	560	680	720	890	1080	1250	1355	1510	1650	1925	2100	2425	2725	3250	640	750	910	1110	1440 g
45	590	720	760	940	1145	1335	1435	1600	1750	2050	2225	2580	2900	3475	675	790	960	1170	1520 g
47	620	760	800	990	1210	1400	1515	1690	1850	2175	2350	2750	3075	3725	705	830	1010	1230	1600 g
49	650	800	840	1040	1275	1475	1595	1780	1950	2300	2500	2900	3250	3950	740	870	1060	1290	1680 g
51	680	840	860	1090	1340	1550	1675	1870	2050	2425	2650	3075	3450	4175	775	910	1110	1350	1760 g

Fillings in Grams for Cooked Sausages Casings, and Spreading Sausages Paste

Ø	175	200	250	300	350	400	450	500	550	600	650	700	750	800	850	900	950 g
43	17	18	20	22	25	28	30	34	36	38	40	43	46	49	52	55	58 cm
50	15	16	18	20	22	24	26	28	30	32	34	36	38	40	42	44	46 cm
55	—	—	17	18.5	20	21.5	23	24.5	26	27.5	29	30.5	32	33.5	35	36.5	38 cm
60	—	—	—	17	18.5	20	21.5	23	24.5	26	27.5	29	30.5	32	33.5	35	36.5 cm
65	—	—	—	16	17.5	19	20.5	22	23	24	25	26.5	27.5	28.5	30	31	32 cm
70	—	—	—	—	16	17.5	19	20	21	22	23	24	25	26	27	28	29 cm
75	—	—	—	—	—	—	—	18.5	20	21	22	23	24	25	26	27	28 cm

continued

TABLE 7.1 (continued)
Sausage—Weight Filling in Grams for Sausages Contents by Size and Diameter for Casings

Ø	175	200	250	300	350	400	450	500	550	600	650	700	750	800	850	900	950 g
80	—	—	—	—	—	—	—	17 [illegible]	19	20	21	22	23	24	25	26	27 cm
90	—	—	—	—	—	—	—	—	—	—	—	—	—	—	23	24	25 cm
105	—	—	—	—	—	—	—	—	—	—	—	—	—	—	20	21	22 cm
Ø	1000	1050	1100	1150	1200	1250	1300	1400	1500	1750	2000	2500	3000	3500	4000	5000 g	
43	61	64	—	—	—	—	—	—	—	—	—	—	—	—	—	—	cm
50	48	50	—	—	—	—	—	—	—	—	—	—	—	—	—	—	cm
55	39.5	41	42.5	44	45.5	47	48.5	—	—	—	—	—	—	—	—	—	cm
60	38	39.5	41	42.5	44	45.5	47	50	53	—	—	—	—	—	—	—	cm
65	33	34	35	36	37	38	39	41	43	—	—	—	—	—	—	—	cm
70	30	31	31	33	34	35	36	38	40	45	50	—	—	—	—	—	cm
75	29	30	31	32	33	34	35	37	39	44	49	—	—	—	—	—	cm
80	28	29	30	31	32	33	34	36	38	43	48	58	68	—	—	—	cm
90	26	26.5	27	27.5	28	28.5	29	30	31	33.5	36	45	51	58	65	78	cm
105	23	23.5	24	24.5	25	25.5	26	27	28	30.5	33.5	42	48	55	62	75	cm

Source: Courtesy J.C. Frentz-encycl.charcuterie-soussana-Document Naturin.

ART OF HANGING SALAMIS WITH TWINE TO
DRY OVER TIME FROM THE CEILING...

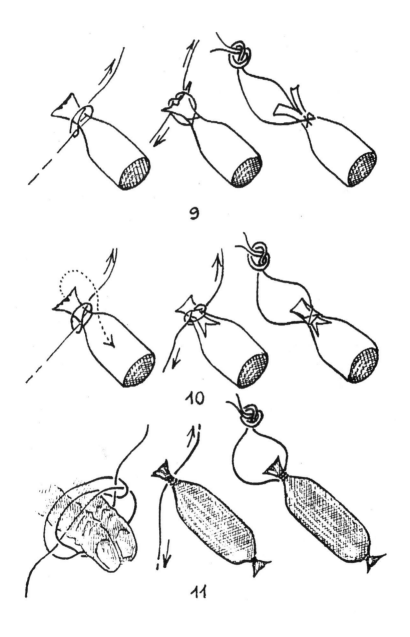

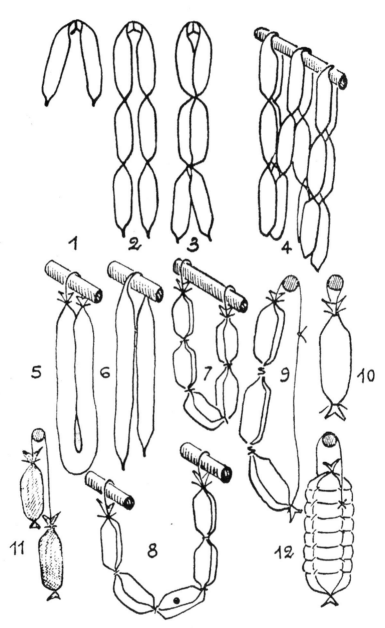

ATTACHES POUR SAUCISSES ET SAUCISSONS

LA CHARCUTERIE EN FRANCE

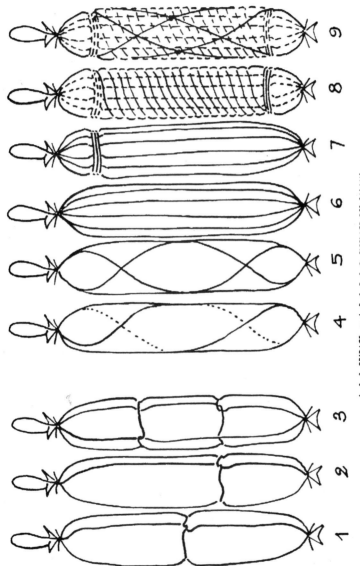

1, 2, 3, BRIDAGE. — 4, 5, 6, 7, 8, 9, FICELAGE DES SALAMI

FIGURE 7.5 BLOOD SAUSAGE WITH CABBAGE.

BLOOD SAUSAGE WITH CABBAGE (BELGIUM)

INGREDIENTS

3 qt. very fresh blood
1 pig head
2.2 lb. cabbage
3 onions
1 lb. pork lard from collar
Pork casings
Salt and freshly ground pepper to taste
Epices marie (mixed spices) or *quatre epices* (see note below)

Chef's note: To make 1 oz. (28 g) of "epices marie:" Mix 5 g nutmeg, 3 g cloves, 2 g basil, 1 g ginger, 2 g fennel, 2 g celery, 3 g laurel, 1.5 g marjoram, 5 g rosemary, 2.5 g thyme, and 1 g savory.

Quatre epices is found in stores in Europe. It actually contains 5 spices: 14% allspice, 20% nutmeg, 32% cinnamon, 18% cloves and 16% caraway.

PREPARATION

Blanch the cabbage in warm water with ½ oz. vinegar for 15–20 minutes. Drain.

Dice the onions.

Melt and clarify the pork lard, and mix it with the onions and cabbage. (This mixture is called saindoux.)

Place the head in a large stockpot and cover with a hearty broth.

Cook until meat is falling off of the bone.

Remove meat and add to the lard, onions and cabbage mixture.

Mix everything together and add seasonings.

Add the fresh blood, which has been previously whipped and strained through a fine mesh colander.

Check seasoning. (Salt and pepper to taste.)

Push mixture into the casings.

Cook sausages in boiling water (with whole savory leaves) at 90°C for 20 minutes, stirring the water gently during cooking.

Gently remove sausages from cooking water and place directly onto a hot grill for a few minutes before serving.

BLOOD SAUSAGE—COUNTRY

INGREDIENTS

½ gal. pig blood, mixed with ½ oz. salt and 1 T vinegar
3 lb. leaf lard
10 lb. diced onions
1 qt. heavy cream
Salt and pepper to taste
1 T epices marie (see chef's note above)
1 T butter
Pork casings

PREPARATION

Remove the skin and dice the leaf lard.

Cook the onions in the butter until soft.

Cool and add to the lard.

Add remaining ingredients, stirring carefully when introducing the blood.

Use a funnel to fill pork casings with the mixture and twist casings to make desired size.

Drop the sausages into boiling water, lowering the heat immediately to prevent bursting.

Using a small pot or ladle, keep stirring the water to keep sausages in motion and cook the sausages for about 15 minutes.

To test for doneness, use a needle or toothpick to pierce one sausage. If no blood emerges, sausages are ready.

Remove sausages from water and place them on a pan lined with a towel or cloth.

FIGURE 7.6   COUNTRY BLOOD SAUSAGE.

FIGURE 7.7

FIGURE 7.8 BOUDIN ANTILLES.

BLOOD SAUSAGES (ANTILLES ISLANDS)

INGREDIENTS

1 lb. pork blood

2 oz. bread (no crust)

10 oz. milk

10 oz. ground meat for sausage

1 whole egg

3 oz. finely chopped chives

1 oz. chopped parsley

⅔ oz. salt

½ tsp. ground white pepper

1 pinch each: ground nutmeg, thyme, and cinnamon

½ oz. rum

1 tsp. cane sugar

Pork casings

PREPARATION

Cut the bread into cubes.

Warm the milk and mix with the bread (panade).

Cook the ground meat with the chives.

Place the bread/milk mixture into a food processor with the cooked meats and process.

Add the blood, eggs, parsley, and seasonings and mix well.

Place mixture over a double boiler (bain-marie) at 40°C/104°F for a few minutes.

Push the mixture into pork casings.

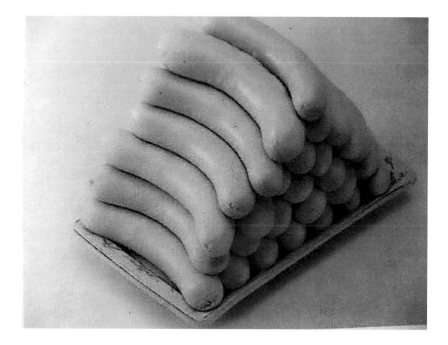

FIGURE 7.9   BOUDIN BLANC.

WHITE SAUSAGE—BOUDIN BLANC

INGREDIENTS

5 lb. pork shoulder or veal (turkey and chicken work as well, but have less flavor)

3 lb. white bread, trimmed of crusts

3 lb. fatback

2 lb. lean pork or veal

25 eggs, beaten

1 pt. white wine port

1 jigger brandy

Salt and white pepper to taste

Pork casings

PREPARATION

Dice the bread.

Dice the pork and the fatback.

Combine all of the ingredients (except casings) and mix thoroughly.

Finely grind mixture with meat grinder.

Fill casings with mixture and twist to form sausages of desired size.

Drop the sausage into boiling water, then immediately lower the heat.

To prevent bursting, use a ladle or small pot to keep water in motion as the sausages simmer (about 20 minutes).

To test for doneness, use a needle or toothpick to pierce one sausage. If no blood emerges, sausages are ready. Sausages will also rise to the surface when ready.

Remove sausages from water and place them on a pan lined with a towel or cloth.

These sausages are best the day they are prepared.

COUNTRY-STYLE SAUSAGE

INGREDIENTS

8 lb. pork (⅓ fat, ⅔ lean)
½ oz. ground white pepper
2¼ oz. salt
⅓ oz. poultry seasoning
12 oz. icy water
Cawl fat fresh as needed

PREPARATION

Grind the meats through a meat grinder set with a coarse plate two
    times.
Add the seasoning.
Divide the mixture into equal portions and wrap each one with caul fat.
Keep them in the refrigerator until ready to use.

SUMMER SAVORY BEER SAUSAGE, FLAMBÉED WITH APRICOT LIQUOR GLAZE

INGREDIENTS

35 lb. Boston pork butt
30 lb. ground turkey
10 lb. dried apricots
3 lb. crimini mushrooms
3 lb. Portobello mushrooms
1 gal. peeled shallots
1 gal. peeled garlic
6 bunches parsley
6 bunches tarragon
6 bunches summer savory
1 lb. Morton quick salt

6 oz. ground black pepper
½ gal. Dijon mustard
100 ft. regular pork casings
1 each 20-oz. bottle Dutch (Jeniver) Gin
1 case (6 bottles) dark ale (such as Chimay)
8 oz. jar beer-soaked green peppercorns
2 L Chardonnay wine

PREPARATION

Grind the meat twice in a meat grinder, using a coarse plate (½″ and ¼″).

Add the salt, pepper, and peppercorns to the meat, as well as the mustard, gin, beer, and wine, and mix well.

Chop the mushrooms, shallots, apricots, and garlic and sauté them in a skillet with the fatback until soft and fragrant.

Chop the herbs and add them to the mushrooms.

Cook briefly, then add mushroom mixture to the meat.

Fill the sausage casings with the mixture.

Keep refrigerated until ready to use.

COUNTRY SAUSAGE WITH
LIVER AND CAUL FAT

INGREDIENTS

1.5 lb. pork shoulder (Boston butt, picnic shoulder)
3 oz. pork liver
6 oz. pork fat
½ oz. salt
1 tsp. ground grey pepper
Pinch of freshly ground nutmeg
1 T chopped parsley
1 oz. dry white wine
Caul fat as needed

PREPARATION

Grind the meats twice in a meat grinder fitted with a coarse plate.
Add the seasonings and the wine.
Divide mixture into equal portions and wrap each portion with caul fat.
Keep sausages refrigerated until ready to use.

FIGURE 7.10   CRÉPINETTES. COUNTRY SAUSAGE WITH LIVER.

CRÉPINETTES—FLAT SAUSAGES

INGREDIENTS

1.5 lb. pork shoulder (Boston butt, picnic shoulder)
6 oz. pork fat
½ oz. salt
1 tsp. ground grey pepper
Pinch of freshly ground nutmeg
1 T chopped parsley
½ oz. dry white wine
Caul fat as needed

PREPARATION

Grind the meats twice in a meat grinder fitted with a coarse plate (½″ then ¼″).

Add the seasonings and the wine.

Divide mixture into 2½ to 3 oz. portions and shape each portion into a ball.

Wrap each portion with pre-washed (and squeezed dry) caul fat. Do not wrap tightly.

Shape each sausage into a flat, rectangular form.

Keep sausages refrigerated until ready to use.

Fry sausages until browned on both sides.

These sausages go well with mashed potatoes.

ITALIAN SAUSAGE

INGREDIENTS

2 lb. 75% lean pork (or ⅔ pork and ⅓ beef)
1 medium onion, finely chopped
1 oz. parsley, finely chopped
1 tsp. crushed garlic
Pinch of nutmeg
1 tsp. paprika
Pinch of oregano
2 oz. toasted, crushed fennel seeds
1 oz. green anise, finely chopped
½ oz. kosher salt
1 tsp. ground black pepper
1 egg
2 oz. cold water
3 oz. red wine
Pork casings as needed

PREPARATION

Grind the meat in a meat grinder fitted with a large hole plate.
Add the parsley, onion, and garlic to the meat.
Add the seasonings, egg, water, and wine to the meat and mix well.
Soften the pork casings in water, with a little bit of vinegar, then stuff
    casings with the sausage mixture.
Refrigerate until ready to use.

FIGURE 7.11   HOT ITALIAN SAUSAGE.

CHEF JACQUES' HOT ITALIAN SAUSAGE

INGREDIENTS

17.5 lb. of pork and veal

Seasoning:

4½ oz. kosher salt	1 tsp. mace
1 oz. black pepper	1 T basil
1 oz. Hungarian paprika	½ oz. chopped fresh Italian parsley
1½ oz. granulated garlic	1 tsp. allspice
1½ oz. granulated onion	8 eggs
2 oz. crushed red pepper	1 lb. milk powder
1 T celery seed	Casings as needed
1 oz. anise seed	

PREPARATION

Cut the meat into 2-in. cubes.

Grind half of the meat using a ½-in. plate.

Grind the other half of the meat using a ¼-in. plate.

Mix seasonings into the meat.

Stuff mixture into casings.

Refrigerate until ready to use.

FIGURE 7.11   PROF. R. FAGNERAY PREPARING LYON SAUSAGE WITH PISTACHIOS.

LIVER SAUSAGE FOR GRILLING

INGREDIENTS

½ lb. pork shoulder meat, cubed

½ lb. throat meat, cubed

1 T kosher salt

1 tsp. quick salt

1 tsp. sugar or honey

Pinch of freshly ground nutmeg

½ lb. of pork fatback

10 oz. pork liver

1 tsp. ground white pepper

2 oz. red wine

5 juniper berries

1 bay leaf

1 thyme leaf

1 garlic clove, crushed

Pork casings as needed

PREPARATION

Place the red wine in a saucepan and add the juniper berries, thyme, bay leaf, and garlic.

Heat briefly to infuse the wine, then let it cool completely and strain.

Coarsely grind the meats in a meat grinder, then add the wine to the meat.

Add the salt, curing salt, spices, and sugar to the meat mixture.

Stuff the mixture into 8-in. long casings.

Suspend sausages in a well-ventilated place at 70°F for 48 hours.

Dry them at 55°F for 8 days.

Slowly smoke sausages for 2 hours, twice—then store refrigerated before grilling.

FRENCH-STYLE GARLIC SAUSAGE

INGREDIENTS

2¼ lb. lean beef

2¼ lb. pork butt (25% fat, 22% lean)

½ oz. chopped parsley

½ lb. of diced Onion

2½ oz. Kosher salt

1 tsp. commercial (6/4) curing salt

1 tsp. ground white pepper

1 oz. finely chopped fresh garlic

4 oz. potato starch

10 oz. cold water

1 tsp. monosodium glutamate (MSG)

1 lb. fresh pork fatback, cut into ⅛ in. dice

Large casings as needed

PREPARATION

Use a buffalo processor with large holes and pass the meat into the triple blades with cold water or ice cubes.

Add the parsley, onion, and garlic and mix together with the meat.

Add the seasonings and the potato starch and mix well.

Soften the casings in warm water mixed with a little bit of vinegar, then stuff the casings with the sausage mixture.

Refrigerate until ready to use.

MERGUEZ I

INGREDIENTS

Lean Beef	40%
(or Beef Mix	60%)
Lamb Mix	40%
Lamb Breast	50%
Beef Fat	20%

Seasonings (g per kg of meat)

	Mediterranean-style	Tunisian	Algerian
Regular salt	22	22	25
Black pepper	3	3	3
Hot pimento	2	2	2
Garlic powder	2	2	2
Kummel	5	—	—
Coriander	—	3	—
Caraway	—	—	4
Red bell pepper	25	40	30
Marjoram	2	—	—
Paprika	2	—	—
Green anise	—	2	3
Oregano	—	2	3

PREPARATION

Blend the spices into the chilled water in order to obtain a purée with a
sour cream-like consistency.

Cube the meat, then grind it in a grinder fitted with a medium plate.

Mix the ground meats with the spice purée.

Stir in the oil.

Soak the casings in water with a bit of lemon juice or vinegar to soften them.

Stuff the sausage into sheep casings.

Refrigerate sausage until ready to use.

MERGUEZ II

INGREDIENTS

18 oz. beef flank rib meat

14 oz. lamb shoulder blade meat

2.5 oz. cold water

1¾ oz. kosher salt

1¾ oz. sweet paprika

1 tsp. ground bird's beak chile pepper

1½ tsp. ground cayenne pepper

1 garlic clove, finely chopped

½ tsp. star anise

½ tsp. oregano

1 oz. olive oil

Sheet casing as needed

PREPARATION: SAME AS ABOVE.

FIGURE 7.12   SPICE MIX FOR CHORIZO OR MERGUEZ.

CHORIZO I

INGREDIENTS: 1 KG EQUAL 2.2 LB.

4 kg pork side lard
3 kg lean beef
3 kg lean pork
200 g kosher salt
40 g grey (mignonette)pepper

15 g cayenne
20–30 g by kg of, red bell pepper
½ oz. red food coloring
28/30 mm pork casings, as needed

PREPARATION

Grind all the meats and the fat (using a ¼″ plate).

Mix the red bell pepper and red food coloring with 2 L of water and process it in a blender to form a purée.

Add the pepper mixture to the meat, then add the spices and mix well.

Push the mixture into 28/39 mm pork casings.

Dry for 12 hours at room temperature.

CHORIZO II

Chorizo sausage originates from Spain and Portugal. It is typically colored and is available in both mild and spicy forms.

INGREDIENTS

5 lb. clean pork side (pig belly)
2.5 lb. very lean pork
2.2 lb. lean beef
2 oz. Kosher salt
½ oz. pepper
½ oz. dry oregano
¼ oz. garlic powder
½ oz. dry red food coloring

2 hot Spanish pimento peppers
1.5 lb. sweet red peppers
3 oz. Olive oil
1 lb. Milk powder
3 oz. red wine rioja.
Pork casings or narrow beef casings as needed

PREPARATION

Grind the beef with an 8 mm plate (larger if it contains tendons).

Mix it with the pork side and lean pork.

Make it thick and to obtain an homogeneous result, use the cutter and sprinkle of milk powder until the grain is soft enough.

Add seasonings. Make a paste with the peppers, oregano, red wine, and olive oil into the cutter and add to the meat paste.

Push sausage mixture into pork casings or narrow beef casings. Style collar.

Let it drip at room temperature for 12 hours.

Stew sausages at 75°F (25°C) for 24 hours.

Dry sausages at 53°F (13/14°C).

Then smoke sausages at 75°F (22–25°C) for 24 hours.

SPANISH CHORIZO

INGREDIENTS

4 lb. 100% lean pork   2 oz. curing salt
2 lb. pork lard   0.5 oz. sugar
20 oz. water

Add a decoction in 3 oz. of red wine rioja of following seasonings:

1 oz. ground black pepper
3.5 oz. sweet Murcia pimento pepper (optional)
1 oz. extra virgin olive oil
3.5 oz. carmine (red food dye)

PREPARATION: SAME AS ABOVE

FIGURE 7.13   CHORIZO SAUSAGES HUNG TO DRY.

FIGURE 7.14   TOULOUSE SAUSAGE.

TOULOUSE SAUSAGE (FRANCE)

Only pork meat is used for this sausage, and 30–40 mm casings are required by law (in France).

INGREDIENTS

2 lb. lean pork meat

2 lb. lean pork shoulder meat

½ lb. hard fat

1.5 oz. regular salt

0.5 oz. pepper

0.2 oz. nutmeg

0.2 oz. allspice

0.5 oz. garlic

0.5 oz. fresh parsley

2 liquid oz. white wine

Pork casings

PREPARATION

Cube the meats and mix them with the fat and the seasonings (salt and pepper, nutmeg, allspice, and garlic).

Chill until very cold.

Grind mixture with a 16 mm plate, then again with a 10 mm plate.

Stir in the parsley and the wine and mix well.

Push the mixture into the casing, and keep sausage in a circular shape (called "Brasse" or "Collier" [collar] in Belgium) when cooking.

PARIS SAUSAGE WITH GARLIC

INGREDIENTS

750 g lean meats, half pork/, half veal
250 g pork collar
18 g sodium nitrite
2 g dextrose
0–5 g nutmeg
2 g grey pepper
4 g garlic
Pork casings as needed

PREPARATION

Grind the meat and the fat with a #8 plate.
Mix with the seasonings.
Push mixture into casings and dry.
Stew sausages until they obtain a nice color.
Smoke sausages to desired taste.

If you like a lighter sausage, replace 18% of the meat mixture with fresh apple purée.

Do the math: 1 oz. = 28.4 g

ALSATIAN-STYLE BRATWURST SAUSAGE

INGREDIENTS

400 g lean pork (or equal parts pork and veal)

200 g throat (skin removed)

220 g fat

150 g chilled milk

30 g spices and additives: *10 g sodium nitrite, 10 g salt, 2 g sugar, 2 g white pepper, 1 g mace, 2 g coriander, 3 g garlic*

PREPARATION

Grind meats and fats with a #8 or #10 plate. Add milk into it through the process.

Add seasoning, mixing well.

Poach or steam 15 min.

Grill few minutes to obtain sear marks, and consume right away.

FIGURE 7.15   BRATWURST WITH RED CABBAGE.

MUNICH-STYLE BRATWURST

Yield: 10 lb. (4540 g)

INGREDIENTS

8 lb. lean veal or cooked veal
    head meat
2 lb. fat
3 oz. salt
0.5 oz. ginger
0.5 oz. ground white pepper
0.5 oz. ground celery seed
0.5 oz. parsley
0.5 oz. ground mace

Rubbed sage to taste
1 lb. cold water or homogenized
    milk
1 g dry lemon seas.
15 g sugar
½ lb. maltodextrin binder #125

PREPARATION

Run meats and fats through the buffalo cutter med speed, add ice or milk chilled (to avoid heating the mix with the cutters).

Add seasonings. Mix well.

Add binder slowly, avoid lumps.

Put into casing.

LYON SAUSAGE WITH PISTACHIOS

INGREDIENTS

750 g lean pork, from ham
250 g fat from neck or throat
20 g pistachios
Sliced truffles (optional)
18–20 g sodium nitrite
2 g dextrose
5 g dairy-based cultures for flavoring (such as Fermento)
1 g nutmeg
2 g grey pepper
30 g cognac
Special casing (baudruche)

FIGURE 7.16   LYON SAUSAGE WITH PISTACHIOS.

PREPARATION

Place the lean meat in salt for 12 hours.

Grind the meat with the fat, using a #8 plate.

Mix meat with seasonings.

Tie the casing off at one end and push meat into the casings.

(If you use truffles, 4 to 5 slices, each.)

Fix the other end of the sausage, hook the sausage and hang to dry.

Stew sausage in an autoclave until desired color is obtained.

DRY SALAMI—SAUCISSON DE MENAGE

INGREDIENTS

Mild brine
1 Garlic clove
2 lb. lean pork
5 oz. lean beef
8 oz. fatback
1+ oz. potato flour (dissolved in a small amount of water)
¼ oz. salt (per pound of meat)
Pinch of curing salt
Pinch of ground pepper
2 drops red food coloring
Pinch of quatre epices*
Casings as needed

*Quatre epices is found in stores in Europe. It actually contains 5 spices: 14% allspice, 20% nutmeg, 32% cinnamon, 18% cloves and 16% caraway.

PREPARATION

Cut beef and pork into 1-in. cubes.
Soak the meat in the mild brine for 3 hours.
Drain and grind meat finely.
Grind fat with a slightly larger plate.
Rub the inside of the bowl with garlic, then place meat and remaining ingredients in the bowl and mix slowly for about a minute, adding the fatback at the end.
Fill casings and tie.
Soak sausages in brine for about 6 hours.
Place in drying room at a constant 60°F temperature until sausages are dried to a desired consistency (about 60–80 days).

(A)

FIGURES 7.17   SOPRESATA, MOUNTAIN SAUSAGES AND SALAMIS TO AIR-DRY.

(B)

Figures 7.17   (continued).

Conversion Table (Weight)

1000 g or 1 kilo	=	2.2046 lb.
1 lb.	=	453.592 g
1 oz.	=	28.3495 g
1 T	=	14.015 g
1 tsp.	=	4.67 g
⅓ tsp.	=	1.55 g

Liquids (Volume)

1 L	=	35 oz.
1 dL (10 cl)	=	3.4 oz.
1 cL	=	0.34 oz.
1 mL	=	0.0034 oz.

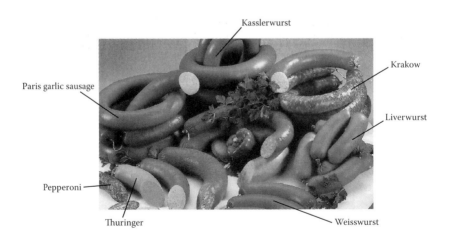

FIGURE 7.18   VARIOUS SAUSAGES.

FIGURE 7.19   STRASBOURG SAUSAGE AND PORK BELLY WEISWURSF WITH SAUERKRAUT.

STRASBOURG SAUSAGE—
SAUCISSES DE STRASBOURG

INGREDIENTS

2 lb. lean beef chuck

1½ lb. fatback

9 oz. ice water

½ oz. kosher salt

½ oz. curing salt

1 tsp. sugar

½ tsp. ground pepper

1 clove garlic, finely chopped

½ tsp. mixed spices

Lamb casings

PREPARATION

Grind the beef and the fat separately, until the ground pieces are the size of a grain of rice.

Place the beef and seasonings in a blender and process while slowly adding the ice water. Incorporate the fat and the chopped garlic.

Mix at medium speed for 1 minute.

Stuff sausage mixture into lamb casings and tie at desired length.

Smoke sausages in a hot smoke box until they have a nice color.

Poach sausages in simmering water until cooked through.

Serve with sauerkraut.

COOKED PEPPERONI

INGREDIENTS

1 lb. lean beef

½ lb. pork fat, cubed

¾ lb. beef

3 oz. chilled water (33°C or 91.4°F)

½ oz. (2½ tsp.) salt

½ tsp. curing salt

½ tsp. sugar

½ tsp. black pepper

½ tsp. paprika

Pinch of coriander

½ tsp. cayenne pepper

1 tsp. crushed garlic

1 tsp. fennel seed

½ tsp. ground anise

Pork casing

PREPARATION

Mix the lean beef with the pork fat and grind it once, using a coarse grind.

Grind the regular beef twice, finely, using a grinding plate with small holes.

Add the water to the finely ground beef, along with the seasonings.

Add the coarsely ground beef and fat mixture to the seasoned beef mixture and mix well.

Push the mixture into the casing.

Store at room temperature for 12 hours.

Cover pepperoni with bouillon or water at 170°F for 25 minutes.

Cool in running water for 5 minutes, then store in refrigerator until ready to serve.

MORTADELLA—MORTADELLE

INGREDIENTS

4 lb. lean pork meat	¹⁄₁₀ oz. (3 g) ground nutmeg
1 lb. pork fat	Pinch of curing salt
5 oz. potato flour	Several drops of red food coloring
Soft brine as needed	½ pt. red wine
1½ oz. salt	3 oz. pistachios
¹⁄₁₀ oz. (3 g) ground black pepper	1 gal. pork or chicken stock
²⁄₅ oz. (11 g) whole peppercorns	Medium beef or large pork casings
¹⁄₁₀ oz. (3 g) sugar	

PREPARATION

Cut the fat into small dice and sprinkle with salt.

Cut the meat into 1-in. cubes and place them in soft brine for 2 hours.

Rinse and strain the meat, then finely grind it.

Place the meat, fat, seasonings, and pistachios (excluding the wine) in a mixer.

Beat slowly for 3 minutes, gradually adding the potato flour while mixing.

Add the wine.

Stuff the sausage mixture into casings.

Place the sausages in a cool place and store for 48 hours.

Smoke moderately for 6 hours.

Cook for 1 hour very gently in the stock

Remove from heat and let sausages cool in the stock for 2 hours.

FIGURE 7.20   HOMEMADE MORTADELLE.

STANDARD MORTADELLE

INGREDIENTS

2 lbs. (900 g) pork (45%)

1.5 lbs. (600 g) beef (30%)

¾ lb. (300 g) ice (15%)

½ lb. (200 g) cubed fat (10%)

1 oz. (28 g) sodium nitrate

0.11 oz. (3 g) sugar

0.11 oz. (3 g) pepper

0.035 oz. (1 g) nutmeg

0.018 oz. (0.5 g) cardamom

0.11 oz. (3 g) phosphate

0.35 oz. (10 g) powdered milk

0.35 oz. (10 g) egg white

PREPARATION

Pre-salt all of the meat 24 hours ahead of time.

Scald the fat slices briefly, just enough to be able to stir them, then chill.

Grind the meats, adding the ice, and then the spices.

Add the cubes of fat and the egg whites to the meat and mix well.

Push the mixture into baudruche casing or 100 mm celluloid casings.

Refrigerate sausage for 24 hours.

Cook sausage in a stew pot or in autoclave for 2–3 hours at 105°F.

Cook for 2 hours at 140°F, then warm smoke the sausages.

Cook sausages in water (or steam sausages) at 170°F.

After cooking, scald the sausages to obtain a smooth, polished casing.

POLISH KIELBASA—SAUCISSON CUIT (COOKED SALAMI)

INGREDIENTS

2 lb. lean beef
8 oz. fatback (medium dice)
2 oz. potato flour, diluted in ice water
¾ oz. salt
1/20 oz. (3 g) curing salt
½ tsp. ground pepper
Pinch of paprika
1 garlic clove, mashed
Beef round casing

PREPARATION

Grind beef very finely.
Place beef in a mixer and add the seasonings (except garlic).
Beat at medium speed, adding the water and potato flour mixture gradually.
Add the garlic.
Add more water if needed.
Stuff the mixture into the beef round casing and tie both ends together, leaving enough string so that sausages can be hung inside of a hot smoke box.
Smoke sausages for about 3 hours.
Place smoked sausages in boiling water, reduce heat, and poach slowly for about 25 minutes.

These sausages are meant to be eaten cold, or slightly warm, with room temperature potato salad.

FIGURE 7.21   KIELBASA.

POLISH-STYLE SAUSAGE—KIELBASA

INGREDIENTS

5 lb. fresh pork butt

1½ oz. (42 g) salt

½ oz. (14 g) sugar

⅓ oz. (9 g) coarsely ground black pepper

1¼ tsp. crushed coriander seeds

¾ oz. (21 g) crushed garlic

4 oz. potato starch

6 oz. very cold water

1 in. pork casing as needed

If sausage is to be cured and smoked, add ⅛ oz. (3.5 g) of commercial 6/4 curing salt to above ingredients.

PREPARATION

Grind the pork very finely.

Place the ground pork in a mixer and add the seasonings (except garlic).

Beat at medium speed, adding the water and potato flour mixture gradually.

Add the garlic.

Add more water if needed.

Stuff the mixture into a large casing and tie both ends together, leaving enough string so that sausage can be hung inside of a hot smoke box.

Smoke sausages for about 3 hours.

Place smoked sausages in boiling water, reduce heat, and poach slowly for about 25 minutes.

These sausages are meant to be eaten cold, or slightly warm, with room temperature dill and vinegar potato salad.

KRAKOW SAUSAGE (MY FAVORITE)

Polish sausage, also called kielbasa in the United States, was originally made with veal shank meat instead of ground beef.

INGREDIENTS

3 lb. veal shank meat, trimmed and cleaned

4 lb. lean pork

3 lb. lard (pork fat)

½ oz. (14 g) mignonette

⅓ oz. (9 g) mixed spices

1 tsp. nutmeg or coriander

½ oz. (14 g) fresh puréed garlic

5 oz. (142 g) maltodextrin or potato starch

3.5 oz. (100 g) tender cure salt

1 oz. sugar

Large pork casing as needed

PREPARATION

Grind the veal shank very finely.

Mix with the seasonings (except garlic) and place it through the grinder a second time.

Add to the bowl of a mixer and beat at medium speed, adding the water and potato flour gradually.

Stir in the garlic and add more water if needed.

Stuff the mixture into large pork casings and tie both ends together.

Leave enough string for hanging sausages in a hot smoke box.

Smoke sausages for about 3 hours.

Place smoked sausages in a steamer and cook slowly for about 20 minutes.

Smoke sausages using wood smoke for 20 minutes.

These sausages should be served with sauerkraut or red cabbage and buttered whole or mashed potatoes.

CHITTERLINGS
SAUSAGE—ANDOUILLETTES

INGREDIENTS

10 lb. pork chitterlings (veal may be substituted)
Bouquet garni
3 onions, peeled and studded with cloves
8 oz. finely chopped shallots
5 oz. dry white wine
3 oz. Dijon mustard
2.5 oz. Salt and pepper to taste
Pork casings

PREPARATION

Rinse the chitterlings in clear water several times.

Blanche chitterlings and rinse again.

Place the chitterlings in a deep ovenproof pot and cover with water, adding the bouquet garni, onions, and salt and pepper.

Cover, bring to a boil, then transfer to the oven to simmer for about 3 hours or until tender.

Cool and cut chitterlings into ½-in. strips.

Sauté the shallots in some butter, add the wine, and cook until wine is reduced by half.

Add the mustard.

Thoroughly mix the shallot mixture with the chitterlings.

Stuff into pork casings and twist stuffed casings to desired lengths.

Poach sausages in simmering water for about 10 minutes, then let cool overnight.

Cook the sausages in a slow oven until golden brown and serve very hot with crispy French fries.

FIGURE 7.22   ANDOUILLETTES.

KASSLER LIVERWURST—
SAUCISSON DE FOIE

INGREDIENTS

1½ lb. pork shoulder meat
1 lb. pork liver
½ lb. fatback
1 lb. cooked tongue (diced)
¼ C. chopped onions
1 oz. chopped truffle peeling (optional)
2 oz. pistachio nuts, blanched and skinned
2 oz. quick cure salt
½ tsp. ground pepper
½ tsp. sugar
½ tsp. epice marie
2 oz. potato flour diluted in white wine
Yellow food coloring
Paper wrapping or large pork casing

PREPARATION

Grind the pork meat, liver, fatback and onions together using a fine
   blade.
Place in a mixer, add seasonings, and beat slowly for about 2 minutes,
   adding potato starch and wine gradually.
Add remaining ingredients (truffles, pistachios).
Roll portions of the mixture up in paper wrapping (or large pork cas-
   ing) and tie each end with butcher's twine.
Poach gently for about 45 minutes.
Remove sausage from wrapping and rub with yellow food coloring.
Smoke in a warm smoke box for about 2 hours.

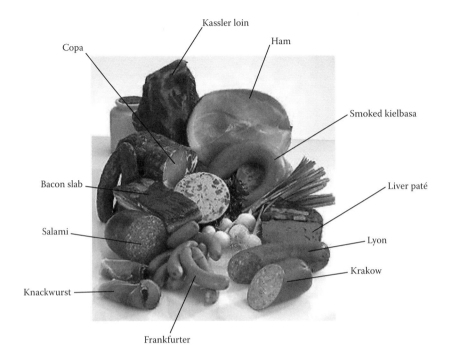

FIGURE 7.23   VARIOUS SAUSAGES.

FRANKFURTERS—
SAUCISSES DE FRANKFORT

INGREDIENTS

2 lb. lean pork	1½ tsp. ground white pepper
3.5 oz. ice water	Pinch of mixed spices
¾ oz. (21 g) salt	1 small of ground cumin
½ oz. of curing salt	Juice of 1 lemon
Pinch sugar	Lamb casing

PREPARATION

Grind meat, salt, and curing salt together.

Refrigerate for 2 hours.

Place the mixture into a blender.

Add the ice water slowly, blending.

Add seasoning and lemon juice and blend well.

Stuff the mixture into lamb casings.

Dry the sausages in a drying room for 3 hours, then cold smoke them for 4 hours.

Poach sausages in water for 20 minutes.

Note: This is the recipe for European-style frankfurters. In the United States, beef is often used instead of pork, or combined with pork. Beef may be substituted for pork in this recipe if desired.

CHAPTER 8

Foie Gras

Foie gras is made from the liver of ducks or geese that have been deliberately overfed, producing fat accumulation in the liver. Migratory birds must gorge themselves in the wild in order store fat (in their liver and elsewhere) to use as energy during their long migration. This practice of fattening geese dates back to Egyptian times, or 3500 BC. A fresco from Saqqara depicts 6 Egyptians force-feeding geese, presumably in order to obtain this delicacy. The practice has continued in various parts of the world ever since.

The Egyptians, Greeks, and Jews feed geese wheat grain, softened with a liquid. The French word "foie" (for liver), like the Italian word "fegato," originates from the Latin term *iecur figatum*, the Roman word for foie gras, (literally "fig-stuffed liver"). The Romans force-fed geese with crushed figs, milk, and honey (Europe did not yet have corn).

Geese were fed this mixture to fatten them after the long journey from Gaul to Rome. Some things don't change: the Gaul's, ancestors of the modern-day French, were already specialists in breeding geese, though the principal areas of production were Boulogne and Calais.

After the Romans, there was not much mention of a foie gras-like dish until the late 18th century, when a version of "paté" appeared 1769 in France.

Jean-Pierre Clause was the chef for the governor of Alsace, Maréchal de Contades. Clause invented a dish called "pâté de Contades," which was a fatty goose liver baked inside a pastry crust. This dish made its way to the table King Louis XVI, who was so impressed that he rewarded the governor with an estate. Later another French chef named Doyen got the idea to add truffles (a specialty of his hometown Perigord) to the famous "pâté de Contades" dish.

FOIE GRAS—FACTS AND
SERVING SUGGESTIONS

Force-fed ducks and geese must be labeled (tagged) by law.

Quantity: There are no specific criteria for determining the quality of foie gras. It's best to use your own judgment, based on smell and appearance.

Weight: The best goose liver weighs between 600–900 g (21–32 oz.). The best duck livers weight between 300–500 g (10.7–17.8 oz.).

Color: Can vary by region from white to yellow to beige or pink. Color should be uniform without any spots.

Texture: Foie gras should be smooth, malleable and unctuous, without any graininess.

Raw foie gras: Uncooked foie gras should be cooked in a skillet over very high heat just long enough give it some color, and seasoned with salt and pepper.

Cooked foie gras: Cooked foie gras can be reheated in a tureen, baked in a pastry crust or in brioche, heated "au torchon," cooked sous vide, etc.

Seasoning and serving: A common practice is to place the livers on a ceramic or enamel dish and season with fine sea salt or kosher salt (15 g/kg), white pepper (1.5 g/kg), heat for 2 hours at 165°F, then refrigerate until ready to use. Reheat for another two hours at 165°F when needed.

Foie gras in a tureen: Cook foie gras in a water bath at 200°F broth until it reaches an internal temperature of 145°F.

Au torchon: Wrap foie gras in cheesecloth and cook in a flavorful broth until it reaches an internal temperature of 145°F.

En brioche: Place the foie gras in the center of the brioche dough. Wrap foie gras with the dough and shape as desired (such as in the form of a mushroom or sausage). Bake at 400°F for 40 minutes.

Baked in a pastry crust: Preheat the oven to 400°F. Wrap foie gras in pastry, leaving a hole in the pastry, then place in the oven and immediately lower temperature to 300°F. Cook until internal temperature reaches 145°F. If the top of the pastry starts to brown, then cover with paper to prevent over browning. Let the pastry sit at room temperature for 2 hours, then fill with a perfumed aspic (piped through hole in pastry crust). Refrigerate.

CHOOSING CANNED FOIE GRAS

Cans labeled with the word "gras" should contain only top grade goose or duck liver. Lower grades are labeled as "parfait," purée, galantine, paté, or mousse. The percentages of liver, spirits, and truffles must be on the label.

Canned foie gras should be chilled before opening, so that it can be removed from the can in one piece.

FIGURE 8.1   RAW FOIE GRAS READY TO BE SEASONED, WITH TRUFFLES NEARBY.

FIGURE 8.2   FOIE GRAS SOUS-VIDE, SERVED OVER CELERY ROOT SALAD REMOULADE AT THE FIRST GALA-DINNER LES DISCIPLES D'ESOFFIER, USA, IN MIAMI 2011.

FIGURE 8.3   PATÉ EN CROÛTE WITH VEAL, DUCK FOIE GRAS, TRUFFLES AND PISTACHIOS.
(COURTESY J.C. FRENTZ.)

FIGURE 8.4   FOIE GRAS EN BRIOCHE. (COURTESY J.C. FRENTZ.)

CHAPTER 9

Terrines and Patés

Terrines and patés are fantastic gourmet dishes that can be prepared from simple ingredients. They are often included on menus as *delices du chef*, or cold buffet *hors d'oeuvres* served before an elaborate meal. They also make a nice light dinner with salad and pickles.

Terrines take their name from the molds that are used to shape them. Meat loaf is an example of a terrine. Terrines can also be prepared with fish or vegetables. Seafood and vegetable terrines are lighter and simpler than traditional meat terrines, and their flavor greatly depends on the freshness of the ingredients used.

Terrines can be prepared with ground meat or with a mixture of ground meat and larger pieces, such as matchstick (julienne) slices of tender meat, pieces of seafood, nuts, etc. Terrines can be prepared in many different flavors, with varied seasonings and ingredients. The meat can be marinated before cooking for extra flavor.

For best results, chill the terrine for several days before serving, as the flavors will develop and ripen.

Paté en croûte is the term for a terrine that is enclosed with dough. Paté de campagne is an exception to this rule—it's called a "paté" but it is actually a terrine (without dough).

PATÉS

General Cooking tips

Preheat oven to 265°F (130°C)
Buée ouverte: "open steam" until paté obtains a light coloration.
Cooking temperature: 212°F (100°C)
Buée fermee: "closed steam"
Cooking duration: longest

Cooking in a humid environment:

Temperature: 85/90°C = 185/195°F
Cooking time: intermediate.

Cooking with steam:

Preheat oven to 90°C = 195°F
Cooking temperature: 85°C = 185°F
Cooking time: short.

It is very difficult to accurately predict cooking times for patés, because every oven is different, and each paté has a different texture and weight. Also, the amount of starch in a paté can affect the cooking time.

The best way to know if a paté is cooked through and has achieved the correct internal temperature is to use a thermometer.

Paté has a tendency to dry out and lose its flavor and a significant percentage of its weight when its interior temperature surpasses 167°F (75°C).

If the paté contains starch, it should cook slowly and for a longer period in order to reach an internal temperature of 75°C to 78°C, which improves conservation of the finished product in crayovac (vacuum).

TABLE 9.1
Paté Seasoning Ingredients
(*Numbers in grams per kilo (Or 0.035 oz. per 2.2 lb). This schema has no limitation.*)

Spice	1	2	3	4	5	6	7
White pepper	2.5	2	2	2.5	2.5	3	2.5
Allspice	1				1.5		
White ginger	0.5			0.3			
Quatre epices	1						
Marjoram					0.5		
Epices fines						6	
Powdered cepes						5–8	
Mace				0.3			0.5
Cardamom							0.3
Nutmeg		1					
Shallots in powder			1.5				
Garlic powder			1				
Laurel			0.2	0.3			
Thyme			0.25	0.3			
Roasted coriander		2					
Sautéed onions	30	30	30	30			
Water orange flower	5	5	5	5			
Alcohol[a]	10	10	10	10			

Note: Ingredient columns 1 to 4 are classic ingredients in spreadable creamy paté. Ingredient columns 5 and 6 are for sliceable liver paté. Ingredient column 7 is the aromatic seasoning for Belgian cream of liver.

[a] In CL (10 CL = 3 oz.)

TABLE 9.2

Special Spice Mixes for Patés (Epices Fines)

Ginger	1 oz.
Mace	4 oz.
Coriander	7.5 oz.
Cinnamon	1.5 oz.
Clove	2 oz.
Bay Leaf	1.5 oz.
Thyme	1.5 oz.
Marjoram	1 oz.
White or Grey Pepper	20 oz.

Note: Grind spices together in a grinder or food processor.

FIGURE 9.1   SEASONING FOR COUNTRY-STYLE PATÉ.

SEASONING FOR COUNTRY-STYLE PATÉS

It's possible to make a very high quality paté by seasoning very fresh pork liver with salt and pepper only. Add some regional flavor with the local seasonings from a particular region or country, such as fresh garlic, shallots or brown roasted onions. Whole parsley is an important ingredient. For alcohol, choose a good quality white wine, port, Madeira, Calvados, Armagnac, etc.

Grams Per Kilos of Mix

Pepper	2–3	2–3	2–3	2–3	2–3	2–3
Thyme	0.20					0.25
Laurel	0–20					
Nutmeg	0.50					
Ginger			0.50			
Mace				0.30		0.50
Marjoram				0.50	0.50	
3 Epices	1.5	1				
Epices Fines			1			

1 k   = 1000 g or 2.2 lb. + 2 g
1 lb   = 454 g
1 oz. = 28.03 g
1 T   = 14.01 g
1 tsp. = 4.67 g
½ tsp. = 2.33 g

RECIPES

COUNTRY PATÉ WITH SHALLOTS AND PEPPERCORN

INGREDIENTS

1 lb. pork liver	½ C. fresh parsley
1 lb., 4 oz. pork throat	½ oz. quick cure salt
½ lb. lean pork meat	½ tsp. ground black pepper
4 eggs	¼ tsp. nutmeg
4 oz. flour	½ tsp. sugar or maple syrup
4 oz. heavy cream	2 oz. dry white wine
1 tsp. fresh garlic	½ oz. whole green peppercorns
½ C. chopped onion	Caul fat

FIGURE 9.2   COUNTRY TUREEN.

PREPARATION

Clean the liver.

Grind liver through a #2 or #3 plate.

Add the finely chopped garlic, onion, shallots and parsley.

Grind the pork fat and then the lean meat through the same plate as the liver.

Add all of the seasonings except the peppercorns.

Mix all ingredients.

Add eggs, flour, and cream, and mix well.

Add wine and green peppercorn.

Place the mixture in a mold preset with fatback.

Cover with caul fat.

Cook until the paté has an internal temperature of 167°F (75°C).

The amount of weight that is lost during cooking depends on the quality of the meats used, but is usually around 8 to 10%.

FIGURE 9.3   TUREENS AND PATÉS DISPLAY.

TRADITIONAL COUNTRY-STYLE PATÉ

INGREDIENTS

2.2 lb. (1 kg) pork liver
4.4 lb. (2 kg) collar fat
2.2 lb. (1 kg) pork throat/neck
2.2 lb. (1 kg) cooked pork skin
17.5 oz. (0.5 kg) whole milk
9 oz. (0.25 kg) eggs
20 g regular salt

3 g freshly ground grey pepper
30 g starch
50–100 g raw onions (varies with taste and region)
1–2 g fresh garlic
Thyme, laurel, and parsley to taste

PREPARATION

This simple recipe is similar to the traditional countryside paté of old. It is a chunky paté, which is first passed through the grinder and then the mixer. All of the ingredients must be extremely fresh. Grind and mix the ingredients the day before you cook the paté, so that it is tastier and more aromatic. The seasonings and the starch are added just at the end of the mixing procedure.

The bottom of this paté must be browned and cooked properly. Start the cooking in a very hot, pre-heated oven, and once some coloration occurs, turn the temperature down and finish it in a humid environment. Cook until the paté has an internal temperature of 167°F (75°C). The amount of weight that is lost during cooking depends on the quality of the meats used, but is usually around 8 to 10%.

When the paté is cooked is cooling, drain it with a liquid aspic made of pigskin.

Note: When you slice the paté, the color is grey because regular salt was used. Chill for two to three days before serving.

CLASSIC COUNTRY-STYLE PATÉ

INGREDIENTS

2 lb. pork liver

4 lb. Pork lard

2 lb. throat or shoulder meat

.5 lb. cooked pork skin

.7 lb. Milk

.75 lb. Whole eggs

20 g/kg regular salt

3 g/kg Ground fresh grey pepper

40 g Starch

50–100 g raw onions

1–2 ea. fresh garlic

Thyme, bay leaves, and parsley to taste

PREPARATION

This recipe is another easy recipe that is similar to the paté made in the countryside in old times, without any special "processing spices."

The paté will be chunky and is first ground then homogenized in the mixer. All the ingredients must be very fresh. The spices and the starch are only added at the end of the mixing. Depending on the quality of the ingredients and the cooking method, the finished product should lose less than 10% of original weight.

After the paté has cooked and cooled, add a very good gelatin liquid "aromatized" to taste.

COUNTRY-STYLE GRIND

Use plaque #8 and cook the paté at 165°F. Warm the oven to start and cook paté in a bain-marie. The starches must be cooked to 165/170°F to avoid fermentation.

FIGURE 9.4 LIVER PATÉ.

LIVER PATÉ—PATÉ DE FOIE

INGREDIENTS

1 lb. pork liver, trimmed
2 lb. fresh skinned pork belly
4 oz. diced onions
3 oz. diced shallots
4 eggs beaten with 4 oz. of flour and ½ pt. of heavy cream
Fat back (for lining molds)
¼ oz. salt
1 pinch curing salt
½ tsp. pepper
¼ tsp. epice marie

PREPARATION

Finely grind all of the ingredients except the egg mixture and seasonings.
Place in a mixer and start beating slowly.
Add the egg mixture and seasonings.
Place mixture in molds and cook in a water bath in a 325° oven until done.
Cool overnight.

Note: This paté is excellent for sandwiches or canapés. It is makes a good substitute for the more expensive foie gras.

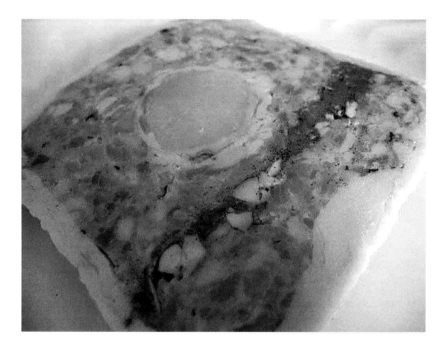

FIGURE 9.5   RABBIT PATÉ DE GARENNE.

WILD RABBIT PATÉ—PATÉ DE GARENNE

INGREDIENTS

2 lb. wild rabbit meat

2 lb. skinned fat pork belly

2 lb. game forcemeat

2 jiggers brandy

4 eggs

4 oz. flour

1 pinch curing salt

1½ oz. salt

3 bay leaves

½ tsp. pepper

½ tsp. epice marie

1 pinch epice marie

1 pinch ground nutmeg

1 pinch thyme

PREPARATION

Cut the rabbit meat and pork belly into small cubes and place in a container with seasonings and brandy.

Mix by hand and marinate overnight.

Place mixture in a mixing bowl and mix at slow speed.

Add the forcemeat, flour, eggs, salt and curing salt.

Proceed as in other paté recipes.

Note: Adding the rabbit's blood to the marinade and making aspic with the bones would enhance the flavor.

DUCK PATÉ—TERRINE DE CANARD

INGREDIENTS

2 lb. boneless duck meat
1 duck liver
2 lb. pork butt
3 eggs
2½ oz. flour
2 oz. truffles
3 oz. Madeira wine (or 2 oz. brandy)
1 oz. salt
Epice marie and pepper to taste
1 pinch curing salt
Fat back for lining terrines

PREPARATION

Cut some of the duck meat into strips and marinate overnight in the wine (or brandy) with half of the seasonings.

In a blender, puree the remaining duck meat, the pork, the eggs, the flour, and the rest of the seasoning.

Line the terrine with the fat back. In the terrine, place a layer of force-meat. Make a roll containing strips of duck, liver, and truffles (held together by some forcemeat). Place this in the center of the terrine, covering with the remaining forcemeat. Cover the terrine with fat back.

Cook in a water bath in a 325° oven until done.

Cool overnight.

Note: Aspic for the paté should be made with duck bones.

FIGURE 9.6   DUCK TUREEN. (COURTESY OF J.C. FRENTZ.)

TERRINE "FORESTIERE" WITH CRANBERRIES OR CHERRIES

INGREDIENTS

2.2 lb. (1 kg) ground pork

5.28 oz. (150 g) "gratin"

10.56 oz. (300 g) mushrooms

3.52 oz. (100 g) shallots

0.70 oz. (20 g) sodium nitrite

0.11 oz. (3 g) dextrose

0.11 oz. (3 g) pepper

0.7 oz. (2 g) nutmeg

0.7 oz. (2 g) epices fines

0.11 oz. (3 g) garlic

5 eggs

2 oz. heavy cream

2 oz. starch or milk powder

2 liquid oz. cognac

2 liquid oz. red wine

2 liquid oz. raspberry vinegar

7 oz. fresh cranberries or cherries

PREPARATION

Marinate the meat with the seasonings, alcohols, a drop of the vinegar and the "gratin."

Grind the meat with a #8 disk/plaque.

Sauté the shallots in a saucepan, then add the mushrooms.

Deglaze the pan with red wine and reduce.

FIGURE 9.7   TUREEN FORESTIERE.

Cool down and add to the marinated meat with the cranberries.

Mix with 5 eggs, 2 g of heavy cream and rich glaze of game meat.

Cook in pre-heated oven until coloration occurs, then add steam and cook until terrine has an internal temperature of 74°C/165°F.

Add rich chicken aspic at end.

FIGURE 9.8   TURKEY TERRINE.

TURKEY TERRINE

INGREDIENTS

35% turkey meat
35% pork throat
30% turkey meat trimmings
Seasonings in g/kg of trimmed meat:
10–20 g (0.35–0.7 oz.) sodium nitrite
3 g (0.11 oz.) dextrose
3 g (0.11 oz.) freshly ground pepper
1 g (0.035 oz.) nutmeg
Armagnac to taste
Time: 12 hours

Seasonings per kg forcemeat:
10–20 g (0.35–0.7 oz.)
   sodium nitrite
3 g (0.11 oz.) dextrose
3 g (0.11 oz.) freshly ground
   pepper
1 g (0.035 oz.) nutmeg
Time of Pre-Salt: 12 hours

Turkey terrine follows:

During the final mixing, add:
Armagnac to taste
1 whole egg
40 g (1.41 oz.) starch
15–20 g (0.53–0.7 oz.) green pepper corns
10% (of whole recipe weight) dried mushrooms
10–15% liver farce (gratin)

Optional: Add 1 g/kg or 0.035 oz. of roast coriander and mushroom juice to taste.

For the mushrooms, use 10 to 20% of the recipe weight and 500 g or 17.6 oz. of "gratin" forcemeat.

Cook the mushrooms whole in a saucepan.

Add a little water with lemon juice (juice of 4 lemons to 5 kg/11 lb. of mushrooms).

Stir and add salt and pepper to taste.

Cover (to retain maximum humidity from the mushrooms (fumet)) and cook on a low burner for 15 to 20 minutes.

Will keep refrigerated up to 15 days.

PREPARATION

Marinate the meats and the trimmings separately in a soft brine cure.

Grind the "farce" with a medium plate.

Place ground meat in a mixer bowl and add the rest of the seasoning and the meat trimmings.

Place the mixture in a terrine lined with thin slices of pork fat.

Cook in the oven at 375°F/195°F, until terrine reaches an internal temperature of 165°F.

Let cool for a half hour, then pour an excellent chicken stock over terrine. (Add 15% of Armagnac to the stock if desired.)

FIGURE 9.9　BURGUNDY-STYLE PARSLIED HAM.

BURGUNDY-STYLE PARSLIED HAM— JAMBON PERSILLE DE BOURGOGNE

INGREDIENTS

1 boiled ham (see recipe), still warm
Red wine vinegar
⅛ oz. finely chopped garlic
1 oz. chopped sour gherkin pickles per pound of meat
1 T chopped parsley

PREPARATION

Remove the skin from the ham without puncturing; use it to line a round, deep stainless steel bowl, with the fat side facing in.

Sprinkle with the garlic, pickles and parsley.

Cut the ham into large chunks, removing most of the fat.

Arrange in layers in the mold until full, sprinkling the chopped mixture between each layer.

Cover with a piece of skin.

Pour a few dashes of vinegar on top.

Add aspic to fill.

Refrigerate overnight; unmold, and cut into regular wedges for services.

FARMER PATÉ WITH GARLIC

INGREDIENTS

1½ lb. pork throat meat
4 oz. pork liver
8 oz. pork breast
1 oz. flour
2 oz. broth or milk
1 whole egg
3 oz. dry white wine
½ oz. salt
1 tsp. chopped garlic
1½ tsp. paté spices
Pork fatback
Caul fat

PREPARATION

Mix all meats and pass them through a grinder plate with large holes.

Add the egg, flour, broth, and wine to the meat, then add the seasonings as soon as the mixture is homogenous.

Put the mixture in a terrine lined with fatback.

Cover it with caul fat.

Put the terrine in a pan with water and cook in an oven preheated to 350°.

Cook until the internal temperature reaches 155°F minimum.

STANDARD PATÉ EN CROÛTE

Chef's note: This recipe includes a special dough called "Paté brisee" for the bottom of the mold and a puff pastry dough for the top and for the meats. We'll use pork and veal with trimmings of raw ham.

INGREDIENTS

300 g pork breast
300 g throat
300 g lean veal
300 g ground shoulder
200 g collar fat
100 g "Gratin"
150 g heavy cream
18 g sodium nitrite
3 g dextrose
3 g pepper
0.5 g coriander
2 g quatre epices
1 egg
Chives, parsley, and chervil to taste
Cognac/Armagnac to taste

PREPARATION

FARCE:

Pre-salt all the meats without the cream.

Pass first through the grinder and then through the cutter along with the eggs, starch and herbs.

SET UP:

Roll out the "paté brisee" into a rectangle and arrange the "crepine" and the smoked pork side slices over the dough, then mix all the meats. Cover first with the smoked lard, then the "crepine" and finally cover with a layer of puff pastry dough, leaving a cheminee (chimney). Decorate as desired. Refrigerate for at least 12 hours.

COOKING:

Like a classic paté en croûte, bake to an internal temperature of 72°C/160°F. Once the paté is out of the oven, use the cheminee to add a bit of cognac or Armagnac as well as coarse, aromatic liquid aspic.

FIGURE 9.10 PATÉ EN CROÛTE.

PATÉ EN CROÛTE

"Stuffing" for Easter Paté (16 in. mold)

INGREDIENTS

1 lb. lean veal shoulder meat
½ lb. pork liver
1½ lb. ground sausage
1 oz. salt
1 tsp. curing salt
1 T sugar
½ oz. truffle peel

FIGURE 9.11A   EASTER PATÉ EN CROÛTE WITH TRUFFLES.

PREPARATION

Line the mold with paté brisee dough ("chemiser" the mold).

Chop the lean meat into ¼ in. cubes and mix with the ground meats.

Season the meat and place into the dough-lined mold. Make sure there are no bubbles and that it is stacked evenly.

Cover with dough and make two chemnées through which the steam can escape.

Decorate as desired.

Cook until it reaches an internal temperature of 160°F.

Let it cool for 1 hour in refrigerator, then pour aspic over paté.

After another hour, pour more aspic over, flavored with an alcohol such as Pineau des Charentes, until it is filled all the way.

Refrigerate for 12 hours before serving.

FIGURE 9.11B PATÉ EN CROÛTE WITH ARMAGNAC.
SAME AS ABOVE, REPLACE MADEIRA BY ARMAGNAC.

DUCK PATÉ WITH PISTACHIOS AND TRUFFLES IN PASTRY— PATÉ DE CANARD EN CROÛTE

INGREDIENTS

7½ lb. duck meat	12 juniper berries
6 lb. chicken and duck liver	16 bay leaves
4 lb. bacon	¼ C. thyme
2½ lb. lean veal	1½ C. paté spice
2½ lb. lean pork	36 egg whites
12½ lb. pork back fat	**Garnishes:**
1 gal. veal consommé	Prosciutto
1 gal. Madeira wine	Duck breast
1 qt. shallots	Smoked tongue
½ lb. mushroom base	Pork fat
½ lb. beef base	Truffles
24 garlic cloves	Pistachio nuts
6 oz. curing salt	Ham

Yield: Nine terrines in standard size loaf pans

PREPARATION

Salt the meats and the condiments and keep refrigerated for 12 hours before processing them in the grinder, passing the through the plates twice.

Refrigerate again for 12 hours, then set the terrine, mixing in cubes of breast, tongue, ham, fat and prosciutto to create a mosaic, and also adding pistachios and julienne of truffle.

SALMON PATÉ EN CROÛTE

Ingredients

700 g salmon fillet
1000 g fillet of sole
3 whole eggs
250 g heavy cream
Salt, cayenne pepper, Noilly (vermouth), and white port to taste
Pastry dough for paté en croûte

Preparation

Poach the salmon for 15 minutes at 175°F (80°C) in a court-bouillon. It should be very easy to remove the bones.

Let salmon cool in the cooking juice.

Mix the sole filets (you can also use turbot) with salt, cayenne, eggs, Noilly, and cream, to make a nice, homogenous pomade.

Check the seasonings.

Line the mold with the pastry dough. Add a first layer of pomade, then a layer of salmon (fillet), then finish filling the mold with the pomade. Close the top with another piece of dough and decorate. (Don't forget the cheminees—chimneys.)

Refrigerate overnight.

Cook paté it in pre-heated oven at 425°F (220°C) for 15 minutes, then lower temperature to 230°F (110°C), until paté reaches an internal temperature of 160°F (65°C).

As the paté is cooling, add good quality fish aspic through the cheminee.

FIGURE 9.12   SALMON EN CROÛTE. (COURTESY J.C. FRENTZ.)

MONKFISH TERRINE WITH
SMOKED SALMON

INGREDIENTS

1400 g monkfish
1000 g sole or turbot fillet
1000 g panade (see below)
500 g heavy cream
8 whole eggs
50 g Cognac/Armagnac
50 g white wine
1 pt. fish "fumet"
40 g green peppercorns

Salt, pepper, and nutmeg to taste
Thin slices of smoked salmon for
   covering terrine
Panade:
600 g milk
200 g butter
250 g sifted flour
Salt, pepper

PREPARATION

Poach the monkfish, then cut into large cubes.
Place in cold "fumet" and heat to 80°C for 5 minutes.

SET UP:

Dice the monkfish into small pieces shaped like thin fingers.
"Barder" (line) the prepared terrine with the thin slices of smoked salmon.
Fill the terrine by alternating layers of "farce" with layers of monkfish.
Cover filling with a piece of crepine and melted butter.

COOKING:

Place the terrine in a bain-marie (water bath) and place in an oven pre-
   heated to 180°.
Cook the terrine until it reaches an internal temperature of 78°C.
Remove from the oven and add a good quality, coarse "fumet."
Press down on the terrines lightly with a weight as they cool.

Sauce:

Prepare 500 g of tomato coulis from fresh tomatoes: peel, seed and sauté tomatoes in a saucepan with butter, minced shallots and a dash of garlic.

Finish by mixing the ingredients in a food processor and adding a squeeze of fresh lemon juice and 250 g of creamy cottage cheese (0% fat), one spoon of chopped parsley, and one spoon of chopped chervil.

FIGURE 9.13 MONKFISH TERRINE.

MARINADE FOR FISH TERRINE

This basic seasoning should be prepared 12 hours in advance. Measurements are in grams per kg (2.2 lb.).

INGREDIENTS

15 g salt

1 g pepper

10 g anise seed

3 g nutmeg

1 g paprika

3 g tarragon

2 g spice mix for fish

Cider vinegar, Noilly (vermouth), white wine, and olive oil to taste

IDEAS FOR FISH TERRINES AND PATÉS:

Lighter terrines and patés are a reflection of modern changes in eating habits and nutrition. Lighter dishes, even with some of the additives, can be as successful as gourmet dishes. Many restaurants add these lighter versions to their menus. Fish terrines, mousses, and patés are, in fact, richer than meat terrines, mousses and patés. They also go well with combinations of vegetables, mushrooms and refined sauces or salads. The most important thing to remember is that the ingredients must be fresh.

Use freshwater fish such as pike, salmon, trout or eel. Saltwater fish such as sole, turbot, John Dory, and snapper are popular with gourmets.

Add truffles or more modest vegetables to produce variations in flavor, from extremely rich to very light! Each and every charcutier, chef, or home cook can find ways to personalize the finished product.

FISH TERRINE WITH "FINES HERBES"

INGREDIENTS

MIX:

1100 g sole or turbot fillet
800 g heavy cream
100 g egg whites
50 g shallots, stirred in butter
36 g salt
3 g pepper
1 g cayenne pepper
Dash of curcuma
Noilly (vermouth) to taste

BLEND:

1000 g cubed salmon fillet
30 g "fines herbes"

SET UP:

Line all sides of a buttered terrine with blanched leaves of Boston let-
tuce, sorrel, or spinach. Fill terrine with the above mix.

COOK:

Bake terrine in a bain-marie (water bath) in a (preheated) oven
at180°F/100°C, until the internal temperature of the terrine reaches
155°F/68°C.

CHAPTER 10

Galantines and Ballotines

GALANTINES

The etymology of the word *galantine* is a bit obscure. Some people believe that it consisted of meat served in aspic. Others believe that the name evolved from an old French word *galine* or *geline,* for hen. They are correct, because traditionally the galantine was made with poultry: chicken or turkey, duck, goose, or pheasant. The meat was deboned, stuffed, and poached in a strong liquid aspic broth, cooled down for two to three days in the refrigerator, then served cold.

Galantines can also be prepared with head cheese or pork paté, including tongue or a stuffed baby pig.

The mosaic appearance of a galantine is similar to terrines or patés en croûte. After cooking, a *nappage* of gelée-aspic is added before decoration. Galantines are always served cold.

Recipes for meat or fish paté existed in the Middle Ages. In the fourteenth century, the chef Guillaume Tirel, alias Taillevent, created a

FIGURE 10.1 GALANTINE EN CHAUD-FROID. (PHOTO COURTESY OF J.C. FRENTZ SEMINAR.)

recipe for paté in which layers of marinated eel filets are alternated with fish stuffing inside a pastry crust.

Rustic terrines, rillettes, and crusts with pork and head cheese have been prepared for centuries in the European countryside, illustrating great ingenuity for using the lesser parts of the freshly killed pig.

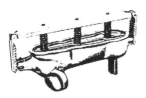

Cochon (oreilles droites)

Cochon (oreilles droites)
moule démontable

Cochon (oreilles couchées)

Dinde

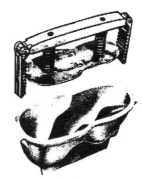

Lapin

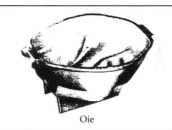

Oie

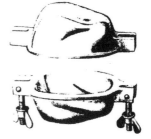

Roulade

Poulet double

FIGURE 10.1b   GALANTINE AND BALLOTINE MOLDS.
(COURTESY OF *ENCYCLOPEDIA OF CHARCUTERY*, J.C. FRENTZ.)

FORCEMEAT FOR GALANTINES

INGREDIENTS

6 oz. (170 g) ½ salt lardons (pre-salted in a soft brine)
1.76 oz. (50 g) carrots, diced and sautéed
2.1 oz. (60 g) white leeks, diced and sautéed
3.5 oz. (100 g) shallots, diced and sautéed
0.88 oz. (25 g) celery, diced and blanched in lemon water
8.8 oz. (250 g) fancy mushrooms, diced
8.8 oz. (250 g) chicken liver, cleaned and washed
1 oz. (30 g) parsley and chervil
1 liquid oz. (3 cl.) Madeira or Port
5.28 oz. (150 g) goose fat
Salt, if needed, to taste
Freshly ground pepper, to taste

PREPARATION

In a pan, melt the goose fat.
Stir in ingredients in the following order: lard, shallots, carrots, white
   leek, celery and mushrooms.
Cook the chicken liver until it is "pink."
Add the parsley and chervil, wine, and salt and pepper, if needed.
Cool down.

USAGE

Add to all stuffing: pork, poultry or game for galantines or special terrines. Use only to 50–250 g per kg (1.76–8.8 oz. per 2.2 lb).

STUFFED BONELESS GOOSE

Same technique as in Porcelet (see ballotines section) with the following ingredients:

INGREDIENTS

1 goose (15–20 lb.)

3 lb. duck or goose meat, cubed

1 lb. ground turkey

18 g sodium nitrite

1 g freshly ground pepper

2 g green peppercorn

30 g pistachios

0.5 g mace

2 ea. eggs

2 g epices fines

"Gratin" made with shallots, celery, onions, diced very thin, and cepes.

Sauté in sauce pan with goose fat for 5 minutes; add pistachios and green peppercorn at least minute.

Deglaze with one ounce of Armagnac.

PREPARATION

Mix all ingredients and chill for two hours.

Stuff the goose with the mixture and wrap the goose with string.

Roast goose slowly at 300°C to start, until it starts to brown.

Turn oven temperature down to 225°C and roast for two hours.

Remove from oven, cool, then add aspic and decorate as desired.

FIGURE 10.2   STUFFED BONELESS GOOSE. (COURTESY OF J.C. FRENTZ.)

FIGURE 10.3   GALANTINE WITH TRUFFLES.

BALLOTINES

Both galantines and ballotines are made from lean pieces of meat such as loins of pork, veal, rabbit, duck, poultry, game, and forcemeat made of ground pork, poultry, and veal. They are examples of fine high quality charcuterie. The proportion of solid meats should be a minimum of 20% by weight. The strips of lard and surrounding aspic must be less than 20% of total weight. Superior quality galantines and ballotines have to include at least 35% of lean and solid meats. Milk, eggs, spices and aromatics are added as binders. Often these preparations are enhanced with a display of truffles, fois gras, loin of principal meat, or forcemeat, arranged in a decorative pattern such as a checkerboard. Classically they are also molded into animals such as ducks, rabbits, chickens, pigs, or they can be molded "au torchon" with a cloth, or reconstituted into the original skin or body of the used animal. They are then called ballotines.

Figure 10.4   Mise en Place (Ingredients) for Ballotine.

FIGURE 10.5   SPICES AND CAWLFAT.

FIGURE 10.6   GALANTINES AND SAVORY BALLOTINES PRESENTATION
TOQUES BLANCHES GALA IN CARACAS-VEN.'92.

FIGURE 10.7 BALLOTINE PREPARATION (5 PHOTOS AS FOLLOW).

Figure 10.8

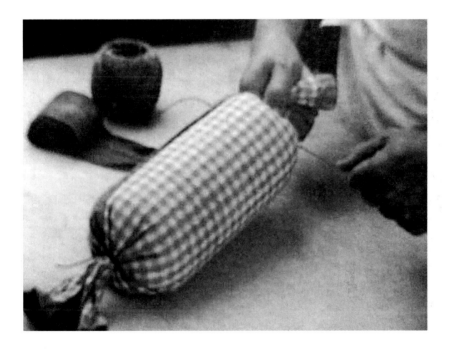

FIGURE 10.9

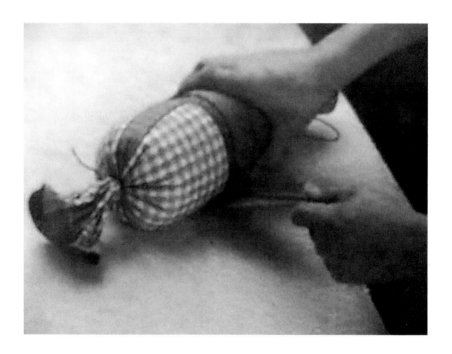

Figure 10.10

FIGURE 10.11 AFTER COOKING INTO A HEARTHY BOUILLON, ASPIC AND DÉCORATE.

BABY PIG BALLOTINE ...

INGREDIENTS

One 15 lb. (6.5 kg) baby pig, uncooked
1.6 kg pork, cubed
1.4 kg ground pork
30 g (per kg) pistachios
20 g sodium nitrite
0.5 g nutmeg
2 g pepper
60 g egg whites
100 g "gratin" with chicken liver
2 g shallots
100 g cepes or morels, soaked
100 g Madeira

PREPARATION

Submerge the whole pig in a soft brine at 1082* beaume for 48 hours.
Rinse and remove all the bones.
Prepare the "gratin" and deglaze pan with Madeira.
Stuff the pig with all of the meats and the "gratin."
Saw the ends and cut the feet nerves/tendons.
Wrap the pig in cheesecloth.
Cook in a broth for about 4 hours, until it reaches an internal temperature of 75°C.

FIGURE 10.12   BABY PIG BALLOTINE. (COURTESY J.C. FRENTZ SEMINAR.)

Chapter 11

More Charcuterie Recipes

PORK RILLETTES

INGREDIENTS

16.75 lb. pork shoulder or spine
11 lb. melted pork lard, clarified
2¼ lb. goose fat, optional
Onions, thyme, bay leaves, garlic cloves, parsley to taste
(For salt, use sodium nitrite, pepper, nutmeg or mace to taste)

PREPARATION

Stir the pork and goose fat in a stock pot on high fire enough to melt
  the fats.
Put the meats in the pot and add spices and condiments.
Cook covered for three hours at low fire and do not mix or move the pot.

Drain the cooked meats and remove the bones, nerves and tendons and crush meats delicately with a fork. This technique will result in a puree.

Clarify the fat, add the juice containing spices and condiments which you'll find in the bottom of the pot, to the meats disposed in a terrine or mold (best in ceramic).

Before complete cool down, add a little bit more of fat as isolation cover to the terrine.

FIGURE 11.1   RILLETTES.

LE MANS–STYLE SPREAD— RILLETTES DU MANS

INGREDIENTS

5 lb. skinless, boneless pork butts, very fatty
Bouquet garni
6 bay leaves
1 pinch thyme
2 onions, clouted with cloves
Salt, pepper to taste

FIGURE 11.2   DUCK LEGS.

PREPARATION

Cut meat in cubes, as for stew.

Place all ingredients in a heavy pot (preferably aluminum).

Add water to almost cover.

Bring to a boil.

Cover and place in a 350°F oven for 2½ hours, or until the meat can be crushed with a fork.

Strain, reserving the juices.

Remove the bouquet garni and bay leaves.

Cool the meat and place in a mixing bowl.

Add back some of the juices, beat at slow speed until the meat is shredded.

Cool the mixture again; place in terrines.

Add fat (part of the drippings if you intend to keep the rillettes stored for a while).

Rillettes can be stored in the refrigerator for a few weeks. They are to be eaten cold as a spread on toasted bread or crackers, or as canapes.

You can also use duck legs.

Marinate them first then cook them slowly in duck fat immersed in a ss/steel pan or clay pot covered, in an oven hot at 250*F for at least 2 hours till the bone turn.

Or cook them till tender, about 45 min, then air dry them or smoke them for duck ham.

BEEF-TRIPES A LA MODE DE CAEN STYLE (CAEN-CITY OF NORMANDY)

INGREDIENTS

3 lb. blanched honeycomb and beef tripe

1 beef foot

2½ lb. onions, diced

3 oz. carrots, sliced

1 leek

2 sprigs parsley

1 pinch thyme

2 bay leaves

1 qt. hard cider (dry chablis could also be used)

2 cloves

3 oz. butter

1½ oz. salt

½ tsp. cafe-ground black pepper

1 pinch cayenne pepper

1 jigger Apple Jack

PREPARATION

Rinse tripe after blanching and cut into 2 in. squares.

Cut beef into four pieces.

Place tripe, carrots and onions in an earthenware casserole with a cover.

Make a bouquet garni with the leeks, parsley, thyme and bay leaves.

Sprinkle with the seasonings, add butter, cider (or wine) and place on fire. Bring to a boil, cover and cook slowly in moderate oven for about 10 hours.

Strain, remove bouquet garni and discard.

SALTED MEAT—"PETIT SALE"

Long ago, before refrigeration, our ancestors, few generations ago, when pork was sacrified for the familial consumption, a good amount of pieces were rubbed with dry salt and spices (dry salt rub as opposed to immerse in "saumure"/brine cure solution) and placed into containers made of stoneware (gres) also called "charniers".

This old traditional technique has almost disappeared, nowadays.

The results were often too salty, with a grayish look, but of course of excellent preservation.

This technique was using only small pieces like a boston bott shoulder 3 to 4 lb. each bone-in or -less., or parts from the flank or bottom of leg called "jambonneau" or shank.

Most of thoses products were served with sauerkraut, lentils, or beans dishes.

INGREDIENTS

4 lb. boston-butt, shank, or pig belly, also called pancetta.

2 oz. of kosher salt

1 tsp. or sodium nitrite

1 tsp. of sugar

2 tsp. of grey ground pepper

1 tsp. of quatre epics

PREPARATION

In a wood board, place the fresh piece of meat and rub with the salt seasoned with other spices energicaly to penetrate into the flesh.

Add a bay leaf and thyme under into the container. Let sit for 2 to 3 days in refrigerator.

Check and flip it for and other day before using.

Or if you don't cook it, use a side of pig belly also called pancetta, roll it and let it cure few days the air-dry it and use it to cook ar as you homemade bacon.

FIGURE 11.3   PETIT SALE.

FIGURE 11.4 PANCETTA.

FEUILLETE OF SALMON EN CROÛTE WITH BABY VEGETABLES

INGREDIENTS

2.5 kg or 4.408 lb. two salmon filets from a whole
1 lb. fresh spinach
1 lb. mushrooms
0.5 lb. baby vegetables
0.5 lb. heavy cream
1 oz. noilly, non-alcoholic anis
Puff pastry dough, as needed
"Crepine," as needed
Butter, salt, pepper, nutmeg, basil, tarragon, to taste

PREPARATION

Trim and clean the salmon filets, boneless and skinless.
Sprinkle them with some Noilly and Anis.
Cover overnight.
Blanch the spinach in some butter until total evaporation.
Add salt, pepper, nutmeg, basil, Noilly and Anis.
Add the cream and reduce.
Sautee the mushrooms in a saucepan with some butter until complete evaporation.
Add salt, pepper and the "fumet."
Reduce until the broth becomes syrupy.

SET-UP FOR EACH FILET:

Over the "crepine," layer as follows: spinach, baby vegetables, mushrooms, one filet of salmon, then mushrooms, baby vegetables and spinach.

Wrap it with the "crepine" and cover with a piece of puff pastry dough.
Cover with a second piece of dough, then pinch the edge and decorate
as you desire.
Brush with eggwash.

COOKING:

After cooling overnight, cook in a 130°F pre-heated oven for 30 minutes.
Serve with a light hollandaise or "beurre blanc."

GRILLED GRAVLAX WITH GREEN PEA PUREE—NORWEGIAN DISH— SERVE APPETIZER SIZE

INGREDIENTS

4 lb. salmon filet with skin on	3 T sugar
Ground black pepper, to taste	3 T aquavit with juniper berries
5 T salt	1 bunch fresh dill

PREPARATION

Marinate the salmon with all ingredients for 24 hours in refrigerator. Press with weight. Turn it twice a day.

GREEN PEA SAUCE

1 C. virgin olive oil	4 lb. frozen early green peas
2 T dijon mustard	½ gal. chicken vegetable stock or fish stock
1 T sugar	
2 T freshly squeezed lemon juice	Kosher salt, pepper, to taste
4 T fresh chopped chives	

PREPARATION

Mix olive oil, Dijon mustard, sugar and lemon juice; set aside.

In a processor or with a hand blender, make a puree with the green peas adding the ½ gal. of stock slowly; strain it in a China cap to eliminate shells and sediments. Add finely chopped chives, salt and pepper while you warm the sauce gently.

To serve with a 2 oz. square of warm gravlax, add a dollop of sour cream and a leaf of fresh dill.

FIGURE 11.5   SALMON GRAVLAX WITH MORELS, ENHANCED BUTTER, AND SALMON CAVIAR.

STUFFED PIGS' FEET WITH TRUFFLES—PIEDS FARCIS

INGREDIENTS

1 lb. cooked, boned pigs' feet
1 lb. cooked, boned snouts
6 oz. strong chicken stock
6 oz. milk
3½ oz. madeira wine or brandy
7 oz. butter
Salt and pepper, to taste
1 oz. chopped truffle peeling
Pork forcemeat
Pigs' veil
Parsley sprigs

PREPARATION

Small dice meat.
Put in a deep pan with seasonings, chicken stock and milk.
Cook slowly until thickened, stirring often.
Add finely chopped truffles and wine.
Spread the mixture evenly to a thickness of ¾ in. to cover the bottom of a shallow pan.
Unmold when completely cooled.
Cut into 2 in. triangles
Make balls with forcemeat.
Flatten the balls and wrap the triangles in the forcemeat, keeping the triangular shape.
Roll in pigs' veil, placing one sprig of parsley atop each triangle before wrapping it.
Broil slowly on both sides until golden brown and serve with mashed potatoes.

ROLLED PORK SHOULDER—
EPAULE ROULEE

INGREDIENTS

1 medium size boneless pork shoulder
2 gal. brine cure
Bouquet garni
2 onions, clouted with four cloves
3 carrots
Fat from patés
Very dry bread crumbs

PREPARATION

Using a special brine pump, inject brine into the meat (especially in thicker parts).

Soak in brine overnight

The next day, after rinsing the shoulder in clear water, roll it in a clean tablecloth.

Tie both ends, not tightly, and place in a deep pan.

Cover with water and bring to a boil. If the water becomes too salty, remove the pork to fresh water.

Add the bouquet garni, onions and carrots.

Bring to a boil; lower heat; simmer about 25 minutes per pound of meat.

Cool off in the bouillon.

The next day, unwrap and wipe dry.

Rub thoroughly with a good paté fat, and roll in very dry bread crumbs.

Note: Rolled pork shoulder is usually eaten cold as a buffet item.

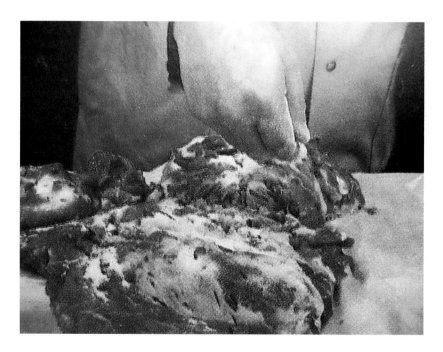

FIGURE 11.6 SEASONING OF THE SHOULDER.

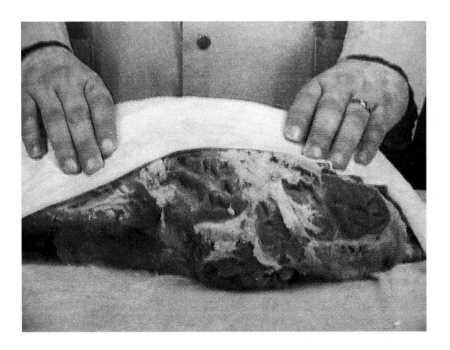

FIGURE 11.7   USE A CHEESE CLOTH TO ROLL.

FIGURE 11.8   WRAP THE PIECE OF MEAT.

FIGURE 11.9   USE A TWINE TO TIE IT TOGETHER.

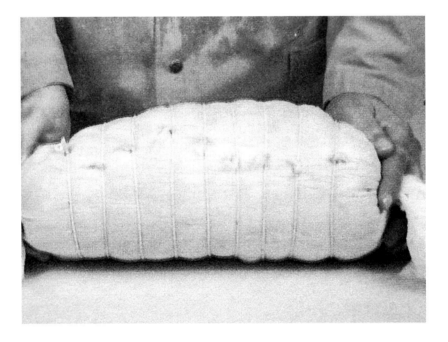

FIGURE 11.10   MORE 3 TIGHT TO COOK IN RICH BOUILLON.

Figure 11.11a

PRESSED HEAD—TETE PRESSEE (HEAD CHEESE)

INGREDIENTS

5 lbs. boneless ears and snouts (2 lbs. ears, 3 lbs. snouts)
2 lbs. pig tongues
2 gal. curing brine
1 bouquet garni
2 pique onions
2 carrots
3 shallots
Chopped parsley
½ tsp. crushed peppercorn
Bouillon
Powered gelatin (1 T per pint of bouillon)
Butter

PREPARATION

Soak meat in curing brine for about 3 hours.

Rinse and place in a deep pan, cover with a good unsalted stock, and add the bouquet garni, onions, carrots and peppercorn.

Bring to a boil and simmer until the meat is very tender, approximately 1½ to 2 hours.

Remove the garniture and skin and bones from the tongues, if any, reserving the bouillon.

Spread a good sized napkin on a table, arranging the meat on it in a cylinder shape, putting the ears next to the cloth and the meat with the tongues in the center.

Cook or sauté the chopped shallots and parsley in some butter and sprinkle over the meat.

Roll into a cylinder; twist both ends carefully.

Place the whole item in stock that has been strengthened with gelatin.

Cool overnight.

Unwrap and slice thinly.

Serve plain or with vinaigrette sauce and thinly sliced onions.

FIGURE 11.11b   BREADED PIGS' SHANKS.

BREADED PIGS' SHANKS— JAMBONNEAUX

INGREDIENTS

Pigs shanks (bone in)   Paté drippings
Brine cure   Browned bread crumbs
Unsalted bouillon

PREPARATION

Use a special brine pump to inject the shanks with the curing brine.

Cook as in boiled ham (see recipe) without cloth wrapping.

When the skin can be easily pierced with a finger, remove from the bouillon.

Remove the bone very carefully; wrap each shank in cloth or sulfurized paper creating a pear shape.

Tighten with butcher's twine, place back in bouillon, and cool overnight.

Remove from bouillon, unwrap, and wipe dry.

Rub shanks all over with good paté drippings and roll in fine, browned bread crumbs.

Slice in wedges before serving as a cold cut.

FIGURE 11.11c   PIGS' FEET IN ASPIC.

PIGS' FEET IN ASPIC—PIEDS EN ASPIC

INGREDIENTS

6 pigs' feet	Bay leaves
1 gal. chicken stock	1 pinch thyme
Bouquet garni	Brine cure
2 clouted onions	

PREPARATION

Clean the feet and shave if necessary.

Split in half, soak one hour in cure.

Tie them back together, using a rectangular piece of stainless steel to separate the two halves, a piece of sulphur paper and some butcher's twine.

Rinse and place in a deep pan with the stock and the rest of the ingredients.

Bring to a boil and cook slowly for about 2½ hours or until soft to the touch.

Cool off in the same stock after removing the vegetables.

Unwrap, sponge dry, place on a rack and brush with a good aspic after decoration.

To be eaten cold with a good vinaigrette sauce.

FIGURE 11.11d   BREADED PIGS FEET.

BREADED PIGS' FEET—PIEDS PANNES

PREPARATION

Proceed exactly as in the recipe for feet in aspic, only after unwrapping the feet, grease them carefully on both sides with a good paté dripping and beat them in bread crumbs.

Cook slowly on charcoal grill until golden brown on both sides.

Serve very hot. It is a delicious entree with French fried potatoes and accompanied by a strong Dijon mustard.

Figure 11.12   Ears and tails.

EARS AND TAILS IN ASPIC—
QUEUES ET OREILLES EN GELEE

INGREDIENTS

Fresh pigs' ears and tails
Bouquet garni
Brine
Unsalted stock

PREPARATION

Shave, clean and cut the tails the desired length.

Clean and shave the ears if necessary.

Soak both in soft brine for 4 hours.

Rinse in clear water, tie the tails and cook in stock with ears and bouquet garni until tender.

Remove bouquet garni; cool.

Strain and arrange ears and tails in a shallow pan.

Cover with a clear aspic made of the bouillon.

Refrigerate, place on rack, decorate, and cover with aspic using a brush.

Serve cold with vinaigrette.

BREADED TAILS AND EARS— QUEUES ET OREILLES PANNEES

PREPARATION

Proceed as in "Ears and Tails in Aspic" recipe, but when cold, rub thoroughly with a good paté dripping and pat gently in bread crumbs. Serve charcoal grilled, very hot, with Dijon mustard and French fried or mashed potatoes.

HEAD CHEESE—FROMAGE DE TETE

INGREDIENTS

5 lbs. boneless ears and snouts (2 lbs. ears, 3 lbs. snouts)
2 gal. curing brine
3 gal. unsalted bouillon
1 bouquet garni
2 clouted onions
2 carrots
½ tsp. crushed peppercorn
Some powdered unflavored gelatin
(1 T per pint of juices)

PREPARATION

Soak meat in curing brine for about 3 hours.

Rinse and place in a deep pan.

Cover with a good unsalted stock; add the bouquet garni, the onions and carrots. The peppercorn can also be added, and the pan should be brought to a boil and simmered until the meat is tender (approximately 1½ to 2 hours).

Remove the garniture, place meat in a strainer, reserve the juice, cool the meat a little and dice it into about ½ in. cubes.

Place in a rectangular mold and strain the juices.

Add gelatin to the liquid; pour into mold to cover the meat.

Cool completely.

Serve sliced (not thinly). Diced pimentos, gherkins, and ham are sometimes added to this dish, but this is a matter of taste. When these items are introduced, it is no longer a *Head Cheese*.

FIGURE 11.13   HEAD CHEESE.

LARDED ROAST PORK IN ASPIC—
ROTI DE PORC PIQUÉ

INGREDIENTS

4 lb. piece ham (center cut if possible)
15 lardons (strips of fatback, the length of the roast)
2 gal. good veal stock
1 oz. salt
1 oz. brown sugar

PREPARATION

Bone and skin meat.

Using a special needle (lardoir), clout the meat regularly with lardons.

Tie up the roast with butcher's twine in regular spaces of ½ in.

Sprinkle the meat with a little sugar (icing) and brown on all sides in a
  frying pan or a very hot oven.

Put the meat in a deep covered pot, cover meat with veal stock and simmer slowly for about 30 minutes per pound, until tender.

Place meat in a deep service bowl.

Cover with an aspic made with the cooking stock, cool overnight and
  slice directly from the bowl.

BOILED HAM—JAMBON BLANC

INGREDIENTS

1 fresh ham, as lean as possible
3 gal. brine
3 gal. unsalted bouillon

PREPARATION

Separate the ham from the shank, putting the shank away for another recipe.

Bone the ham carefully by sliding the bone out and lift the skin almost entirely in one piece using a very sharp knife.

Trim most of the fat.

Inject cure with special needle in all parts of the ham and soak the ham for 3 to 4 hours in the brine cure.

Rinse the ham in clear water, roll in an appropriate size piece of cloth, tie both ends, then the whole ham, starting from one end to the other, as to make a perfect cylinder, but not too tight.

Place ham in deep pan, cover with stock, bring to a boil and simmer very slowly for about 4 hours to 20 minutes per pound of meat.

Remove from fire, unwrap and wrap again, tightening up the strings.

Place ham back in stock and cool off completely before unwrapping and slicing.

CHAPTER 12

The Art of Pickling
Preserving Foods with Vinegar and Alcohol

Pickling is a technique that is used to preserve all kinds of foods, including vegetables, meat, fish, eggs, fruit and even nuts. In addition to prolonging the shelf life of the food, pickling adds flavor and can transform a fresh food into an entirely new product (such as cucumbers into gherkins). Some other examples include vegetables in a sauce, vegetables preserved in olive oil, or fruits preserved in alcohol. It's important to remember that pickled foods are only as good as the ingredients with which they are made.

FIGURE 12.1   INGREDIENTS NEEDED FOR PICKLING.

BASIC CORNICHONS
(PICKLED CUCUMBERS)

Pickling is a simple operation, requiring only two steps.

To make cornichons or pickled gherkins, you begin with small gherkin cucumbers.

Wash them in cold water then thoroughly dry them with a towel.

Add a layer of coarse pure salt to a large bowl.

Place a single layer of the cucumbers on top of the salt and cover them with another layer of salt.

Repeat with remaining cucumbers, making layers of salt and cucumbers.

Let them rest in the salt for 24 hours.

Rinse the cucumbers in a large bowl of cold water with a few drops of vinegar to remove the excess salt.

FIGURE 12.2

Dry cucumbers with a clean towel. Now they are ready to go into their containers.

Sterilize glass jars by placing them boiling in boiling water for few minutes.

Put the cucumbers into the jars, adding aromatic dry herbs such as thyme, laurel, tarragon, dill, condiments like allspice, mustard seeds, fennel seeds, coriander seeds, whole pepper, whole mini inions, crushed peeled garlic, etc.

FIGURE 12.3   MAKING CORNICHONS: WASHING THE CUCUMBERS.

Choose high quality vinegar that contains at least 4–6% acetic acid, in order to obtain the best preservation. Fill the jars with vinegar.

The fastest method is to boil the vinegar first so that it will penetrate faster into the cucumbers.

Seal the jars using a sterile technique and store them.

After 3 to 4 weeks, the pickles will be ready to taste.

Always use a wooden spoon or wooden tongs to remove pickles from the jar.

FIGURE 12.4   MAKING CORNICHONS: LAYERING THE CUCUMBERS IN SALT.

FIGURE 12.5 MAKING CORNICHONS: RINSING SALT FROM CUCUMBERS.

TIPS FOR SUCCESS

Do not use iodized salt (table salt) for pickling. It contains anti caking ingredients and will make the brine cloudy, as well as discolor the ingredients. Choose instead a pure salt, sea salt, kosher salt, or pickling salt.

Brown malt and red wine apple cider bring more flavors to the food but can also alter the color of certain vegetables.

Always respect recipes, and never skimp on the designated quantities of vinegar or alcohol.

Make sure your containers are well sealed and stored in a fresh place.

FIGURE 12.6   MAKING CORNICHONS: DRYING CUCUMBERS.

FIGURE 12.7   MAKING CORNICHONS: ADDING VINEGAR AND HERBS TO JARS.

FIGURE 12.8   USING WOODEN TONGS TO REMOVE FINISHED CORNICHONS FROM JAR.

FIGURE 12.9   VARIANTE COLORFULL.

FIGURE 12.10   PICKLING VEGETABLES WITH COOKING SAUCE.

PICKLES WITH SAUCE

Let's go to a step further and add a sauce. When you cook vegetables in a vinegary sauce or in a sauce with added alcohol, the cooking liquid can become the means of preservation. Later you can enjoy both the vegetables and their sauce. You can even incorporate mustard, which makes a thicker sauce and goes well with many other ingredients. (In Europe, this condiment is known as piccalilli sauce.)

Depending on the aromatics you choose and the cooking time, the pickles can be either sweet or sour and can be either soft or all dente.

Alcohol acts just like vinegar, perfuming and preserving the food. Beer and wine do not have enough alcohol content on their own, but spirits like like Cognac, Armagnac, and Calvados work well and add flavor to the preserved food. The flavor of the liquor is also enhanced by the aromatic qualities of the fruits and vegetables used.

FIGURE 12.11   PICKLING VEGETABLES WITH COOKING SAUCE.

FIGURE 12.12   PICKLING VEGETABLES WITH COOKING SAUCE.

FIGURE 12.13 PICKLING VEGETABLES WITH COOKING SAUCE.

MORE PICKLING POINTERS

Both vinegar and alcohol preserve food by penetrating the food and replacing the natural liquids. Alcohol is the most effective, because it actually destroys microorganisms, while the acetic acid in the vinegar only stops their proliferation. Both alcohol and vinegar work by osmosis to add flavor to pickled foods.

Wine vinegars are the finest, but cider vinegar, malt or distilled, are good and less expensive substitutes. Choose the highest quality vinegar possible, as it is such an essential ingredient.

In pickling, like all culinary endeavors, starting with the best and freshest ingredients produces the best results.

Vinegar reacts with most metals, so make sure you use stainless steel, glass, enamel, or stoneware containers when working with vinegar.

Be meticulous with sterilization procedures. Botulism is a serious disease.

Bon appetit!

FIGURE 12.14   ITALIANS PRESERVE THEIR CONDIMENTS WITH GOOD OLIVE OIL AND VINEGAR AND ARRANGE THE VEGETABLES IN COLORFUL LAYERS.

CHAPTER 13

Photos of Charcuteries Displays for All Occasions...

This is an extra gift to you... my new "disciple."

By buying this book, you are now able to produce all of my recipes...I give you some advise or more guidance to display them on platters, trays or mirrors for your brunches, banquets, receptions, VIP parties, or also any competition projects that you are working on to reach another level of culinary excellence.

Always keep a centerpiece as the focus of your idea and work around it with fine, small accompaniement pieces, such as pickles or colorfull mini vegetable baskets.

Anything become possible if you decide this morning to breathe well and set your mind to it. The horizon is infinite.

There are now following...

FIGURE 13.1   HEAD CHEESE, TURKEY BREAST, BRESAOLA, LIVER MOUSSE, MARQUESA WITH HAM AND SALAMI.

FIGURE 13.2   ASPIC TUREENS WITH VEGETABLES AND PICKLED GARNISHES.

FIGURE 13.3   VARIOUS PATÉS AT CARRIBEAN FEAST IN MARACAIBO, VENEZUELA.

FIGURE 13.4   MORE PATÉS AT CARRIBEAN FEAST.

FIGURE 13.5   CHAUD FROID, TURKEY, AND BAKED HAM IN ASPIC.

FIGURE 13.6 BEEF WELLINGTON, SMOKED PORK LOIN, BRESAOLA, AND SALAMI.

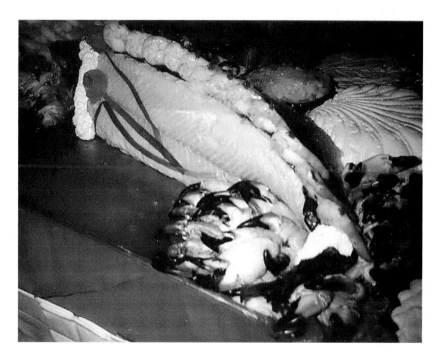

FIGURE 13.7 STUFFED BONELESS KING SALMON WITH FLORIDA STONE CRAB CLAWS.

FIGURE 13.8 MOUSSE FOIE GRAS, COPA, PROSCUITTO, AND VEAL BREAST.

FIGURE 13.9 BRESAOLA, HEAD CHEESE, JAMBON BLANC, AND SMOKED DUCK BREAST.

FIGURE 13.10   COLD CUTS FOR BRNCH.

FIGURE 13.11   PATÉS MOUSSES.

FIGURE 13.12   MARQUISE DRESSED WITH HAM, COPA, BRESAOLA, AND RIVER OF ROAST BEEF.

FIGURE 13.13   CHICKEN MOUSSE WITH MORELS AND BROCCOLI.

FIGURE 13.14   ASPIC OF SALMON.

FIGURE 13.15   GALANTINES AND BALLOTINES FOR BRUNCH IN CARACAS.

FIGURE 13.16   TWO BONELESS STUFFED KING SALMON, U 10 PRAWNS, AND STONE CRAB CLAWS AT VIP RECEPTION IN MIAMI.

FIGURE 13.17 A WEDDING TRAY OF DELICACIES.

FIGURE 13.18   A YOUNG CHEF JACQUES.

FIGURE 13.19   LA MARIPOSA (BUTTERFLY) OF COLD CUTS IN VENEZUELA, 1990.

FIGURE 13.20   STUFFED VEAL LOIN WITH SQUASHES AND PICKLED VEGETABLES BOUQUET.

FIGURE 13.21   A SALMON BELLEVUE.

FIGURE 13.22   TRILOGY OF PATÉS: BOUILLABAISSE, RABBIT ASPIC, AND
FOIE GRAS IN ARTICHOKE BOTTOM WITH BRIOCHE.

FIGURE 13.23   CLASS AT JOHNSON AND WALES UNIVERSITY IN MIAMI, 1995.

FIGURE 13.24   CLASS AT JOHNSON AND WALES UNIVERSITY IN MIAMI, 1995.

FIGURE 13.25   MY FRIEND PAUL PRUDHOMME HELPING AT TEAM USA PRACTICE, 1994.

FIGURE 13.26   BASTOGNE BONE-IN JAMBON IN CRUST.

FIGURE 13.27   SMOKED RABBIT NAVARIN WITH WILDBERRIES IN SMOKY MOUNTAINS, 2006.

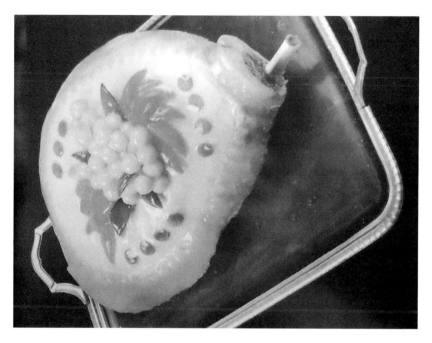

FIGURE 13.28   JAMBON WITH MIRABELLE ASPIC. (COURTESY OF J.C. FRENTZ SEMINAR.)

FIGURE 13.29   ZAMBONI—VEAL FOOT STUFFED WITH PISTACHIOS FARCE FINE.
(COURTESY OF J.C. FRENTZ SEMINAR.)

FIGURE 13.30   AN AGING ROOM FOR PROSCUITTOS IN PARMA, ITALY.

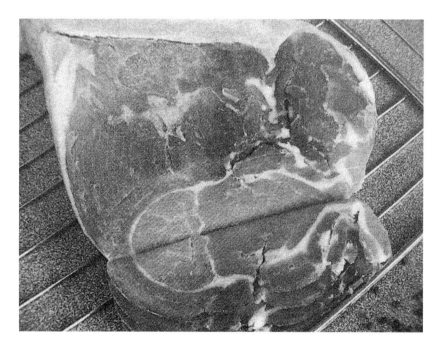

FIGURE 13.31   AIR-DRIED HAM, PARMA STYLE.

FIGURE 13.32   DRY BONELESS HAM FROM ESPELETTE/BAYONNE REGION, RUBBED WITH ESPELETTE HOT PEPPERS AND OLIVE OIL, SERVED WITH FRESH FIGS.

APPENDIX

INVENTORY OF WEIGHTS

1. Weight of raw ham, total .... 150 kg
2. Weight of raw ham, after trimming .... 142 kg
3. Weight of raw ham, injection .... 160.4 kg
4. Weight of raw ham, brine or curing .... 157.5 kg
5. Weight of raw ham, bone and skins .... 15 kg
6. Weight of raw ham, trimmings cured .... 30.50 kg
7. Weight of raw ham, meats molded .... 112 kg
8. Weight of raw ham, after cooking and unmolding .... 97.0 kg

CONTROL AND CALCULATION OF COSTS

9. Percent of Injection
10. Weight with Brine
11. Weight of Trimmings, Cured, and Cleaned of Salty Brine
12. Weight of Meats
13. % Brine during Cooking
14. Weight of Trimmings Cured, but Clean of Brine or Salmur
15. Weight of Meat
16. Rendering Meats
17. % of Cooking Waste
18. Rendering Through Process
19. Product Total

Duration of Storage for Refrigerated Animal Meat

	Temperature	Packaging	Time of Holding	Observations
Meats				
Beef: carcasses	4	None	10 to 14 days	
Beef: carcasses	−1.5 to 0	None	3 to 5 weeks	
Beef: carcasses (boneless)	−1.5 to 0	None, with 10% CO_2	9 weeks	
Beef: slabs, boneless	−1.5 to 0	Crayovac	12 weeks	
Beef: pieces	4	Plastic Wrap	1 to 4 days	
Retail Cuts				
Retail: pieces	4	Crayovac	14 days	
Retails: pieces	2		9 to 12 days	80% O_2+; 20% CO_2
Ground Beef	4	Plastic Wrap	24 hours	

Duration of Storage for Refrigerated Animal Meat (continued)

	Temperature	Packaging	Time of Holding	Observations
Ground Beef	4	Crayovac	7 to 14 days	
Ground Beef	2		3 to 5 days	80% O_2+;j 20% CO
Pork				
Carcass	4	None	8 days	
Carcass	−1.5 to 0	None	3 weeks	
Slabs	−1.5 to 0	Crayovac	3 weeks	
Retail Pieces	4	Plastic Wrap	3 days	
Ground	4	Plastic Wrap	24 hours	
Slices of Charcuterie Products	4		3 to 6 weeks	Depending on water retention and bacterial quality
Wiltshire Bacon	4	None	3 to 5 weeks	

continued

Duration of Storage for Refrigerated Animal Meat (continued)

	Temperature	Packaging	Time of Holding	Observations
Lamb				
Lamb	4	None	1 to 2 weeks	
Lamb	−1.5 to 0	None	3 to 4 weeks	
Sheep	−1.5 to 0	Crayovac	10 weeks	
Veal	4	None	6 to 8 days	
Veal	−1.5 to 0	None	3 weeks	
Organs	−1.5 to 0	None	7 days	
Fowl				
Fowl	4	Plastic Wrap	7 days	
Fowl	−1 to 0	Plastic Wrap	2 weeks	
Fowl	−2.2 to −1	Plastic Wrap	3 weeks	
Fowl	−2	Plastic Wrap	3 to 4 weeks	

Note: Humidity of 85 to 95% needed normally unless usage of special packing like Crayovac or other.

LIGATURES FOR DRYING
AND HANGING SAUSAGES

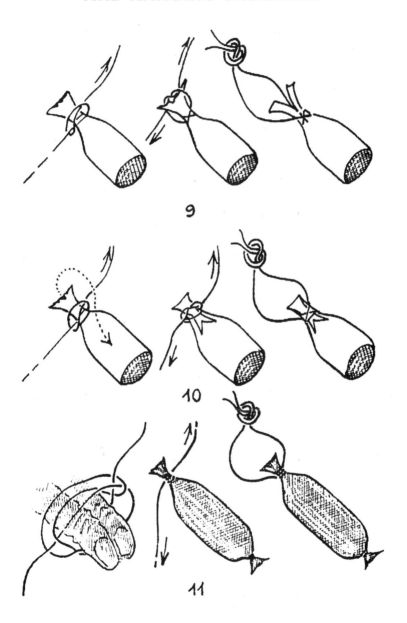

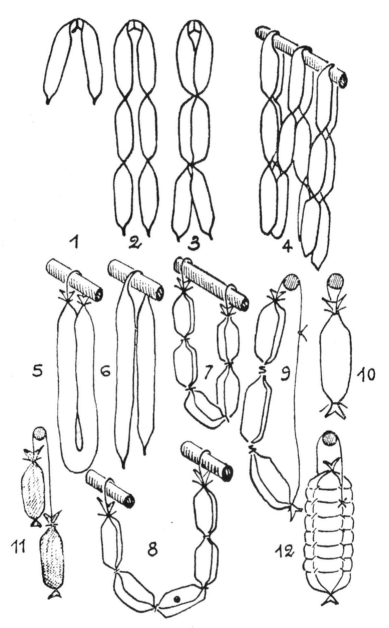

ATTACHES POUR SAUCISSES ET SAUCISSONS

LA CHARCUTERIE EN FRANCE

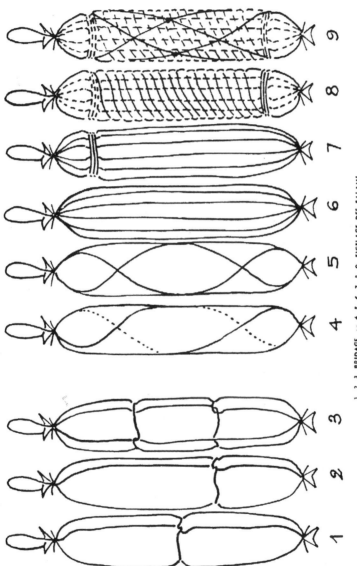

1, 2, 3, BRIDAGE. — 4, 5, 6, 7, 8, 9, FICELLAGE DES SALAMI

INDEX